Luminos is the Open Access monograph publishing program from UC Press. Luminos provides a framework for preserving and reinvigorating monograph publishing for the future and increases the reach and visibility of important scholarly work. Titles published in the UC Press Luminos model are published with the same high standards for selection, peer review, production, and marketing as those in our traditional program. www.luminosoa.org

The publisher and the University of California Press Foundation gratefully acknowledge the generous support of the Richard and Harriett Gold Endowment Fund in Arts and Humanities.

Building Green

Building Green

Environmental Architects and the Struggle for Sustainability in Mumbai

Anne Rademacher

UNIVERSITY OF CALIFORNIA PRESS

University of California Press, one of the most distinguished university presses in the United States, enriches lives around the world by advancing scholarship in the humanities, social sciences, and natural sciences. Its activities are supported by the UC Press Foundation and by philanthropic contributions from individuals and institutions. For more information, visit www.ucpress.edu.

University of California Press
Oakland, California

Suggested citation: Rademacher, A. *Building Green: Environmental Architects and the Struggle for Sustainability in Mumbai.* Oakland: University of California Press, 2018. doi: https://doi.org/10.1525/luminos.42

Library of Congress Cataloging-in-Publication Data

Names: Rademacher, Anne, author.
Title: Building green : environmental architects and the struggle for sustainability in Mumbai / Anne Rademacher.
Description: Oakland, California : University of California Press, [2018] | Includes bibliographical references and index.
Identifiers: LCCN 2017044981| ISBN 9780520296008 (pbk.) | ISBN 9780520968721 (e-edition)
Subjects: LCSH: Sustainable architecture—India—Mumbai. | Architects—India—Mumbai. | Architecture—Environmental aspects—India—Mumbai. | Urban ecology (Sociology)—India—Mumbai. | Rachana Sansad (College). Institute of Environmental Architecture.
Classification: LCC NA2542.35 .R335 2017 | DDC 720/.470954792—dc23
LC record available at https://lccn.loc.gov/2017044981

CONTENTS

ILLUSTRATIONS

TRACING ENVIRONMENTAL PROCESSES:
CONNECTING PLACES, SOCIAL AGENTS, AND
MATERIAL FORMS

How does an anthropologist focused on environmental and political change in Nepal come to study among environmental architects in Mumbai?

One of my most constant, and constantly fascinating, groups of interlocutors in Kathmandu was an extraordinarily committed and effective set of workers for the non-governmental organization called Lumanti. Tireless in their advocacy, and fearless in the face of repeated official threats and obstacles, I was fascinated by the group's tenacity and effectiveness. But I also noticed that part of its strength derived from connections to a robust network of housing advocacy groups across South Asia. Among the most prominent members of this group was the Society for the Promotion of Area Resource Centers, or SPARC, and the network of organizations that made up Slum Dwellers International. SPARC's central office was in Mumbai, and so, expecting to further my understanding of South Asia's regional urban housing politics, I traveled there for the first time in 2008.

A few weeks into that first stay in Mumbai, I received a call from the head of the Rachana Sansad Institute of Environmental Architecture. We had never met, and I was, until then, unaware that RSIEA existed. The institute head invited me to deliver a lecture to environmental architecture graduate students on the subject of urban ecology. My first response was a confused hesitation. What, I wondered, did architects have to learn from an environmental anthropologist? However, in part out of sheer curiosity about how this community of architects—a group with which I had not previously had research contact, and a field in which I had no formal training—would engage with a lecture on urban ecology delivered from the perspective of someone trained in environmental sciences and sociocultural anthropology, I accepted.

Continuing my conversation with the head of the institute, I quickly learned that RSIEA was the first architecture program in India to offer a formal master's level degree program in environmental architecture. It had pioneered what has since become a widely replicated training model throughout the country, adapted in some places with a heavier emphasis on theory, and in others with a more intensive focus on professional praxis.

As we discussed the Institute and its mission, it became clear to me that the form of "environmental architecture" codified through the creation of this formal degree program, and made up of specific and selected content, was a potentially important arena for understanding urban ecology in practice in a guise I'd not previously considered. It suggested the potential to challenge my longstanding focus on marginalized groups and marginal urban landscapes by considering how ideas and practices of nature are made among a very differently positioned group of social actors, professionals seeking to balance ecological and social well being through design. The relationship between the built form of slum housing and environmental politics had occupied my analytical attention for over a decade, but I understood little about how power and wealth asymmetries figured among professionals caught between those making policy and those who commissioned and controlled the making of the formal built landscape. My optic into coupled political and environmental transformation thus shifted from informal and marginalized housing to the ways that the makers of the formal built landscape imagined and enacted an alternative eco-political urban future. In the process, I found the distinction between the formal and informal built landscape to be, at best, a heuristic.

The present project connects to my previous research through its central theoretical and analytical questions, but the histories of Kathmandu and Mumbai are quite distinct, separate, and unique. They undergird dramatically different social and biophysical settings within which to undertake any study of the social life of urban environmental sustainability. At the same time, the connective flows of information, ideas, and affinities that brought these locations together in my field research experience—as nodes in a housing advocacy network that brought together Kathmandu and Mumbai-based rights activists—were real and significant. Specific relations of power were formed and reinforced as interconnected local organizations worked to address their cities' housing and environmental dilemmas, forms of power we stand to miss if we stop at the conceptual boundary of two distinctive, separate cities in two countries with wholly distinctive histories.

Nevertheless, Nepal's capital city, Kathmandu, has a long and layered history as a trading center of many kingdoms; it remained on the outskirts of colonial empire. Mumbai (earlier Bombay) is quite roundly a colonial city, and its fort, white and native enclaves, slums, and suburbs have distinctive qualities even as they compose patterns that one might also see in other modern Indian ports and presidency cities that were forged in the colonial encounter with the British. As Gyan Prakash writes, "the physical form of Mumbai invites reflection on its colonial origin . . . in fact, the Island City occupies land stolen from the sea," and it "bears the marks of its colonial birth and development."[1] Unlike Kathmandu before the tragic earthquakes of April 2015, Mumbai's built environment has few monuments to a deep past, yet it testifies to land reclamation and occupation in the construction of a vast empire of colonial commerce.[2]

To recall its past as built on land "stolen from the sea" also invites consideration of the Anthropocene future, in which the entire Indian subcontinent is cast, first and foremost, in a sea sure to "steal" coastal zones afresh.[3] But the coming dynamics of sea level rise and transformed water access patterns in Mumbai and across South Asia form only one cluster of the many questions that bridge matters of ecosystem ecology to the contemporary making of this city that was first rendered through land filling, concretization, and encroachment. Mumbai is many islands fused into one; its present coastal, littoral, and intertidal ecosystem dynamics are that transformation's legacy.

Arguably, the ecological ruptures through which contemporary Mumbai was made over the past one and a half centuries were, at the time of my fieldwork, more dramatic than those that had shaped Kathmandu. But as two of the fastest growing metropolitan centers in the region in the later part of the twentieth century, Kathmandu and Mumbai experienced similar conditions as well. With the project at hand anchored to Mumbai, then, my challenge was in part to bring a legacy of tracing political-ecological connections between two South Asian cities to a grounded investigation of the unique ecological, historical, and social context of environmental architecture in Mumbai. It was also to move from an optic on the social experience of informal housing and slum advocacy to a formal and professional world of practicing urban architects. It is this endeavor that I undertake in *Building Green*.

· · ·

Learning a new city is neither easy nor automatic, and a single lifetime is hardly sufficient to become fully acquainted with any city's layers. I first arrived in Mumbai dependent on the care and guidance of others, and many years later I remain a student of its vast and constantly changing ecosocial landscape. The project that informs this book would have been impossible without the generous and vibrant intellectual and social worlds that opened for me a welcoming space, and that invited me to learn, teach, and dwell among a group of urban professionals committed to an alternative vision for the city's future.

I am deeply grateful to the students, faculty, and administrators of Rachana Sansad Institute of Environmental Architecture for their extraordinary warmth, consistent collaborative support, and endless intellectual gifts. I worked among them as an anthropologist with keen interest (but no prior training) in architecture, and this in itself could have been rightly regarded as burdensome at best, boldly reckless at worst. Yet the faculty and students received my presence among them in quite the opposite spirit: they embraced the perspective and background I could contribute, and they patiently shared their own. My respect for this community of teachers, learners, and practitioners has only deepened with time, and it is my sincere hope that the content of this book honors their unbounded gifts of time, insight, and powerful, determined aspiration. I have assigned pseudonyms to all of the student-architects who appear in *Building Green,* but as very public figures, most faculty members are named. I must emphasize here that this study, the analysis, and the core arguments I advance are my own. So too, are any errors that remain in the text.

While in Mumbai, an intricate web of intellectual and personal support gave me the critical input and restorative energy I needed to complete this work. I am deeply grateful to the Anand Family, Nikhil McKay Anand, Ramah McKay Anand, Shaina Anand, Roshni and Abraham Yehuda, Bharati Chaturvedi, Brinda Chugani, Urvashi Devidayal, Rohit Tote, Kapil Gupta, Devika Mahadevan, Amita Baviskar, Bharati Chaturvedi, Aban Marker Kabraji, Khojeste Mistree, Priya Jhaveri, Dr. C.S. Lattoo, Shilpa Phadke, Arjun Appadurai, Shekhar Krishnan, Maura Finkelstein, Jonathan Shapiro Anjaria, Harris Solomon, and Ar. Sharukh Mistry. Ar. Mishkat Ahmed provided essential research assistance as I conducted the survey work for this study; her creative energy and thorough engagement with this project breathed unusual life into quantitative data collection and management.

Several organizations provided the research support that made field work for this project possible. I am grateful to the American Institute for Indian Studies, Tata Institute of Social Sciences, Max Planck Institute, and Partners for Urban Knowledge Action & Research for their assistance in Mumbai. In addition to other forms of support, colleagues at TISS generously provided much-needed office space for reflection, interview transcription, and writing. Midway through a significant period of fieldwork, I received a New York University Global Research Institute Grant, which afforded me a productive period to write while in residence at NYU-Berlin. There, Gabriella Etmektsoglou, Roland Pietsch, Nina Selzer, Sigi and Almut Sliwinski, Susanah Stoessel, Carmen Bartl-Schmekel, and Miruna Werkmeister welcomed me into their worlds, and often their homes as well. My preliminary analytical work on this project was challenged and strengthened through deep engagement and thoughtful critique from colleagues at L'Ecole des Hautes Etudes en Sciences Sociales in Paris, where I was appointed as a Visiting Fellow. I am especially grateful for instructive guidance from Francis Zimmermann, Miriam

Ticktin, Loraine Kennedy, Blandine Ripert, and members of the Center for South Asian Studies at EHESS.

As has long been the case, the inspiring and highly original scholars who contribute to the Ecologies of Urbanism in Asia Network provided continuous support for this work through their fascinating case studies, theoretical interventions, and warm collegiality. My partner in this enterprise, K. Sivaramakrishnan, continues to model the best possible combination of impeccable scholar, inspiring collaborator, and generous friend. Together we are grateful to the Hong Kong Institute for the Humanities and Social Sciences, whose unwavering support of our quest to better understand the diverse ecologies of urbanism across Asia has enabled multiple fruitful projects and fostered generative connections between scholars across Asia, Europe, and North America.

In New York, my colleagues at the NYU Institute for Public Knowledge, particularly Caitlin Zaloom, Eric Klinenberg, and Gordon Douglas, provided much of the support necessary to turn fieldwork and analysis into a finished book. Many versions of many chapters in *Building Green* trace their origins to the IPK Library; the entire book was sharpened through an early peer review made possible by generous support from IPK. Parts of this book were further enriched by meticulous comments from colleagues who took part in the Urban Beyond Measure Symposium at Stanford University, the Rutgers University Human Ecology group, the Urban Landscape Studies Group at Dumbarton Oaks, the Critical Perspectives on Urban Infrastructure Workshop at University College London, and my colleagues in NYU's Departments of Environmental Studies, Anthropology, and Social and Cultural Analysis. The support of departmental chairs in those units—Terry Harrison, Susan Anton, Dale Jamieson, Peder Anker, and Carolyn Dinshaw—was essential throughout many phases of field research and writing.

In specific places and moments, colleagues and friends offered support, critical input, or simply care that helped to bring this project from research to analysis to a finished book. I wish to thank in particular Peder Anker, Gustavo Azhena, Manu Bhagavan, Neil Brenner, Mary Cadenasso, Vanesa Castán Broto, Kizzy Charles-Guzman, Sienna Craig, Arlene Dávila, Nina Edwards, Julie Elman, Henrik Ernston, Tejaswini Ganti, Asher Ghertner, Gokce Gunel, Jeanne Haffner, Erik Harms, Karen Holmberg, Maria Ivanova, Natasha Iskander, Sophia Kalantzakos, Richard J. Karty, Mary Killilea, Liz Koslov, Andrew Mathews, Cindy McNulty, Pam McElwee, Mariana Mogilevich, Harvey Molotch, Laura Murray, Laura Ogden, Sara Pesek, Salman Quereshi, Christina Schwenkel, Tamara Sears, Maria Uriarte, Tyler Volk, and Austin Zeiderman. I am equally indebted to dear friends, whose unwavering personal support kept me centered and sustained. Barbara and Roger Adams, Steve Curtis, David Elman, Sasha Gritsinin, David Heiser, Christopher Hoadley, Ko Kuwabara, Momo Holmberg Tang, Tamara Rademacher, Rana Rosen, Kai Schafft, Stephanie Steiker, and Kenny Tang offered warm fellowship at critical moments in the long journey from fieldwork to book. I also gratefully acknowledge the assistance of Evelyn Baert in the final stages of editing this manuscript, and Ayaka Habu in developing all aspects of the NYUrban Greening Lab Initiative that was born alongside this work.

My longstanding academic mentors remain my greatest source of intellectual inspiration. James Fisher, Michael Dove, Helen Siu, and the community of scholars in Yale's Dovelab and Agrarian Studies Seminar continue to nourish my work with their critical guidance, continued encouragement, and precious gifts of time. As I was writing *Building Green,* we mourned the passing of Eleanor Zelliot, my most cherished academic mentor, who in the formative role of my undergraduate advisor became a source of lifelong inspiration. Her vibrant presence is missed every day.

My parents, Ronald and Nancy Rademacher, are the loving beginning of all I do and am. This book reaches back to them with gratitude and with reverence.

In the middle of this project, I experienced a life-changing health crisis. Surviving, and healing in its aftermath, depended on the unconditional, and in many ways miraculous, generosity of friends and family. I am forever grateful to Leah Mayor, Gil Mayor, Jacque Haloubka, and Jordan Mayor for creating a healing space in their home, and a nurturing place in their family, as I inched my way back to the life and health I am so lucky to have regained. This book is dedicated, with love and incalculable gratitude, to them.

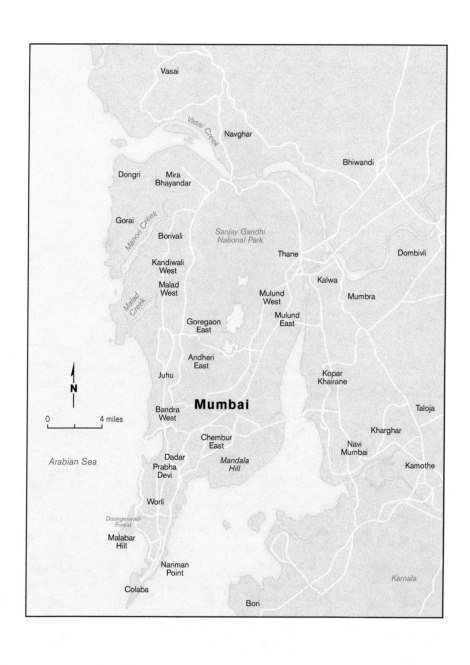

City Ascending, City Imploding

I.

One by one, we filed back into a rickety van. Days of travel over smooth highways and potholed lanes, narrated in hours of conversation, song, laughter, and silence, had fostered the distinctive familiarity that sometimes develops with time shared in transit. A few days into the journey, we'd fallen into territorial patterns: by will or by default, we'd claimed and repeatedly reoccupied a specific seat in the van. Sliding into my place, I joined in a collective, exhausted exhale. Our energy was spent; our senses were full.

From the early morning hours, our group of thirteen architects and professors had been touring the headquarters, and then several building sites, of the Bangalore property development firm called Biodiversity Conservation India (Ltd.). We'd covered the firm's philosophical basis for environmental design, learned a set of technical strategies for achieving building efficiency and maximizing environmental performance, and, then, finally, we walked several construction sites to experience some BCIL projects in the making.

The van revved its engine, filling the air with the sour sweetness of exhaust. Tired but still curious, I thumbed through the day's collection of brochures, pamphlets, and fliers. Settling on a BCIL brochure for potential clients, I skimmed the introductory pages. "DON'T JUST BUY A HOME, BUY INTO A CAUSE," urged its opening page, the text laid out in capital letters over a large green exclamation point. "This is the future of urban living," it continued, "Welcome aboard."

My eyes raced over descriptions of the many residential projects that were planned or underway at BCIL. We'd walked several of those project sites over the course of the day, and I'd found each more impressive, innovative, and surprising

FIGURE 1. Construction in Mumbai, 2012. *Photo by the author.*

than the last. The architects I traveled among—all practicing professionals who had returned to graduate school to enroll in a master's degree program in environmental architecture—were noticeably inspired; each site seemed to present something new to marvel at. The residential developments bridged the ideas we'd spoken of at our orientation at BCIL headquarters and the *things,* the buildings, the ideas rendered in material form—or, at least, the things in the midst of *becoming* material form.

My attention switched from the brochures to an announcement delivered from the front of the van, where the Head of Rachana Sansad Institute of Environmental Architecture, Roshni Udyavar Yehuda, stood balanced precariously against the sway and bounce of an uneven dirt road. "There is one more stop," she declared, "and it will include lunch!" Exhaustion gave way to excitement and relief; we were ready to eat.

Closing the brochure in my lap, I paused for a moment over its concluding text:

BCIL. If you haven't heard of us before, that's alright. We like it that way. Because we believe that some of the most revolutionary ideas in the world have quiet, unknown beginnings. BCIL is all about tomorrow's thinking today. The very fact that you're reading this page is important; it tells us you're thinking on the same lines. It tells us you are exactly the kind of person we're looking for: the kind that looks ahead, sees all the angles, and sees holistic understanding. Our emphasis on community and conservation is not an alternative. It is an imperative for the future.

With that, the van pulled to a dust-choked stop. Renewed by the thought of lunch, we filed out into the searing sun, following Yehuda as she guided us to a small building that appeared to be a private residence. A small sign read, "Alternative Technology Foundation"; the group's founder greeted us warmly at the door.

Our plates soon heavy with dal and rice, we settled on scattered cushions to eat and chat. After a few minutes, our host offered a more formal greeting, followed by a short lecture about the work of the ATF. As his talk came to a close, its tone grew urgent. He said:

> People like us—architects and designers committed to green design—see the future. Common people do not; my neighbor does not. The future is our responsibility. We are like soldiers of sustainability. You are all like soldiers of sustainability.

This was a day like many others in the Rachana Sansad Institute of Environmental Architecture, a day spent seeking to learn about environmental, or "green," design from encounters with specific examples. In the classroom and on field trips such as this one, we sought the philosophical basis for what we studied, a kind of architecture that was very new, and yet distinctly ancient, all at once. We sought demonstrations of its material possibility and the technical strategies that made it plausible. Most importantly, however, it was a day of reinforcing the idea that, left on their present trajectory, India's cities would suffer severe social and environmental crises. Eventually, conditions would become so extreme that a new vanguard of urban professionals who could navigate the terrain of sustainability—let us call them green experts—would be needed to lead those cities to remedy, and to a salvaged future.

The "soldiers of sustainability" I studied among were but one part of this essential vanguard, yet they regarded their work as central to its mission. In the near future, their capacity to think in an integrated way, and to imagine and design future built forms that would embody BCIL's "holistic understanding," would be nothing short of essential; the same propensity to "look ahead" and "see all the angles" could eventually form the very basis for human urban survival.

Perhaps most importantly, tomorrow's environmental architects were cultivating a shared sense of belonging to and being among this vanguard. Our sense of good and right design was cultivated together in the context of our training; it left us with a shared moral ecology foundational to the salvaging of the future city—indeed, to the salvaging of the very future itself.[1]

The long day behind us, and our appetites now quiet, we returned to the van for the last segment of the day's journey. We rode in near-silence, soon crossing onto smooth pavement that seemed to lull each head to sleep. I clutched again my stack of BCIL brochures, mindful of the day's crossings of past, present, and future. Recalling the words spoken at ATF, my mind echoed with phrases, "people like us" and "seeing the future." In the span of a day and a daylong journey, the architects I traveled and studied among were a step closer to joining the vanguard of a multivalent, global social movement called urban sustainability.

II.

As if to satirize twentieth century categories that located Mumbai in the "developing" world, a highly visible, citywide advertising campaign for the real estate development firm India Bulls proclaimed *India Bulls: Consider it Developed*. I first encountered this slogan in 2007, a time when Mumbai was alive with construction. Across the city, large, flimsy walls marked the temporary boundaries between the city standing and the city under construction. Behind the barriers rose the hidden components of the future city, sunk in vast pits that secured their foundations. From the roadside, one could only rehearse the omnipresent slogan: *Consider it Developed.*

A layered, massive mosaic of urban material and social life, Mumbai in that period was palpably transforming in real time. Optimism reigned in amplified public spheres alive with celebratory spectacles, media coverage, and forecasts of seemingly endless economic growth. The city was emboldened in large measure by India's relative insulation from an otherwise debilitating global financial recession; it seemed positioned to pronounce its place at the nerve center of an undeniably ascendant Asia. Such a place meant little on the global economic landscape if not that India, and its financial hub, Mumbai, were unquestionably "developed."

Yet the Mumbai of that particular present was also mired in almost iconic poverty; the city's buildingscape was famously dominated by slum housing, and transected by notoriously substandard transportation, electricity, and water delivery infrastructure.[2] In that moment, Mumbai was a complex historical product of colonial spatial production, often-opaque and brutal politics, and sometimes spectacular scandal, each driven as much by bureaucratic authority and corporate power as by India's oft-referenced status as the world's largest democracy.[3,4]

Globally prevalent mappings of urbanization, in which Mumbai regularly figured as a major location on a "planet of slums," circulated as they did, but the city nevertheless rode a wave of growth, however asymmetrical, through which developers and government officials promised a Mumbai yet to come.[5,6] "Consider it developed" conveyed more than the enormous capacity for growth and change that the building industry celebrated in its everyday construction spectacles; it also captured a defiant postcolonial confidence. Mumbai was a city whose time had come, emblematic of a euphoric Indian century. At least, perhaps, the slogan allowed one to revel in that possibility. *Consider* it developed, because in the twenty-first century this is not only reasonable, it is also wise. It would be difficult to dispute that the city was a good investment.[7]

International reports and government ministries outlined an Indian future animated by dizzying rates of change. As the National Planning Commission called for an almost seven percent increase in energy production to keep up with projections of *nine percent* growth, the consulting firm McKinsey Global Institute predicted an astonishing expansion of Mumbai's built landscape.[8] The city's commercial built-up area alone, it claimed, would grow from 2.9 billion square feet in 2005 to 20 billion square feet by 2030.[9] Just a few years later, in 2014, the global real estate firm Cushman & Wakefield reported that net office space across eight major Indian cities had increased by sixteen percent in the first half of the year, compared to the same period the previous year.[10] This was to say nothing of the residential and housing sectors, in which growth and transformation drove countless policy studies and notoriously lucrative speculative markets.[11]

Beyond the vexing socioeconomic challenge of the present, then, stood the shining promise of growth. Those who could participate in that growth enjoyed tremendous power and watched their personal wealth multiply. In this context, developers, builders, and financiers enjoyed a special status. But equally important, if not always as powerful, were certain urban planners, architects, and urban policy professionals, who, in visible if not always overtly powerful ways, voiced reminders that the city faced environmental challenges as well. They sometimes championed, and sometimes contested, official pronouncements about the appropriate path to the Mumbai in the making.

If the grueling poverty and vulnerability that characterized city life for most Mumbaikars could in some ways be assuaged by euphoric narratives of economic growth, the city's biophysical future was that hope's undoing. Studies of the present and possible effects of environmental degradation, frequent and erratic major storm events, loss of coastal land to sea level rise, possibly catastrophic flooding, and, ironically, sub-continental water scarcity all punctuated predictions for the city's future ecological reality.[12] Mumbai's energy and food security scenarios, its water budget, air quality, and vulnerability to storms would all reshape the biophysical stage for the city's unfolding.[13] New and sometimes massive populations of migrants were expected to mobilize in response to coastal conditions and sea level changes across South Asia; this would rework the human landscape as it reshaped the urban interface between the city and the sea.[14]

Dire poverty and future environmental stresses thus held the promise of growth in uncertain suspension: the idealized key would be to grow in a way that maximized ecological vitality as well as economic profits, and that effected more equitable distribution of a vast array of socioenvironmental benefits. To achieve this, a particularly "green" expertise was essential: one that could guide the form of the new city toward environmental and social adaptability.[15]

III.

In February of 2012, on a first walk through the middle class residential neighborhood that would be my fieldwork home that year, I spotted a newsstand display
hung with the attention-nabbing covers of the day's latest papers and magazines.
The week's issue of *Time Out Mumbai* was on prominent display, beckoning its
readers with an enormous headline: "Imagine Mumbai." Intrigued by the premise, and by the cover image of a pleasant, tree-lined coastal urban promenade, I
bought the issue and tucked it among my things. Later, working my way through
the magazine's articles and images, I paused over the many examples of a supposedly possible future version of Mumbai. This Mumbai was laden with lush urban
landscapes woven of leafy parks, diverse open spaces devoted to leisure, and vegetated zones devoted to the unique ecology of a healthier, more climate-resilient
coastline. Each example was at once profoundly unfamiliar, and yet—or so the
convincing renderings suggested—profoundly possible for those who dared to
"imagine Mumbai."

The portal to this barely recognizable city—communicated in this form to a
small, elite, and relatively young readership—based its declaration of timeliness
and possibility on the particular bureaucratic moment. According to formalized
urban planning cycles, Mumbai had ostensibly—though not exactly in practice—
created a new urban development plan in twenty-year intervals. Since 1966, the
Maharashtra Region and Town Planning Act, established in that year, required
every municipal corporation to prepare and implement city development plans.
Calls to an elite young readership to *Imagine Mumbai* echoed the task of professional urban planning publics as they debated the appropriate form and content of
the current urban plan's successor. In this sense, the future plan for Mumbai could
be treated as an open question, ripe for certain publics to reimagine.[16]

By this time, the official plan-making process was already controversial, in part
because a consortium that included French consulting firms had been appointed
to write the new plan.[17] In part as a response, prominent calls for public participation (and spectacles that sought to enable it) created a sense among a specific
subset of elite and professional Mumbaikars that their individual and collective
acts of "imagining Mumbai" mattered, and moreover, that they could and should
be galvanized to influence the form and content of the new urban plan.

Broadly ecological sensibilities dominated the public meetings and exhibitions through which these publics sought to influence the plan's form. The idea
of "open space," a category encompassing calls for more recreational and leisure
space, concerns about public health and well-being, and a host of ecological conservation objectives, came to capture and convey a complex of potential remedies
for the spatial and environmental deficiencies of the present and the biophysical
challenges that climate change ensured. In this sense, calls to integrate urban sustainability concerns into the new development plan assumed the form of a civic

imperative.[18] To promote the conditions needed to constitute a fuller, more ideal Mumbai was to promote an attendant biophysical and material form.

The *political* issue of how, precisely, to ensure that an amended development plan would be both formulated *and* operationalized, however, was usually tucked into subtext. This had the effect of foregrounding the planners, architects, and other urban professionals who imagined, narrated, and justified it rather than the political economic structures, bureaucracies, and circumstances that enabled or prevented them. The urgency of the moment, emboldened by the looming plan deadline but already evident in intensifying concern over the risks to coastal cities posed by climate change, seemed to excuse the discursive circumvention of the political mechanics of actual change. Mumbai's complex and famously problematic bureaucratic, corporate, and development apparatus would have to be reformed, but precisely how was muted, if even present, in calls to bring urban sustainability into the new plan.[19]

But back in the pages of the bourgeois print media voice of *Time Out Mumbai,* a young, elite, English-speaking readership was nevertheless called on to take responsibility for the Mumbai of tomorrow. The most effective way to do that, it suggested, was through design thinking. "How would *you* redesign the city?" one article asked, suggesting an enticing combination of agentive power and civic duty. Each piece proposed a different strategy or focal Mumbai geography, but all converged on a single point: "open spaces." Vegetated, accessible public areas designed with a combination of trees, gardens, and leisure in mind, were a primary tool for achieving a more desirable city, according to this logic.

To "imagine" Mumbai in this context was thus an invitation to rethink its socio-ecological destiny, its spatial configuration, and the patterns that came together in the process of urban development. It was also a confident gesture that seemed to imply that such rethinking could itself have real, material consequences. The exercise was a first step that, if mobilized in design arenas that could spark unspecified collective civic agency, might change the spatial course of Mumbai's future. Placing the work of transforming imagination into action in the hands of urban professionals, then, bestowed a sense that they could have an influence—or at least a voice—in ensuring Mumbai's very survival.

The historical moment was clear and indisputably urgent. True ecological and economic vitality were still possible for Mumbai, but the opening signaled by the development plan was finite and pressing. It was in this moment, in this complex and dynamic city, that I embarked on an ethnographic journey among architects seeking training in green design.

. . .

In this book, I take special interest in the training, thinking, and voices of a particular group of Mumbai-based architects. Theoretically, architects were among the urban professionals potentially poised to envision, convey, and create a material bridge between the city's present built landscape and the environmental, material, and social city still to come. Architects were among those working to design and actualize the estimated 2.3 billion square meters of floor space that would, in fewer than two decades, rise across India.[20] In Mumbai, a subset of architects aspired to do this in a way sensitive to the altered energy, water, and environmental vulnerability profile of the entire subcontinent, while trying to also address the deep socioeconomic asymmetries that demanded change.[21] Like architects and planners from many points across history and place, their design aspirations sometimes linked to imagining new social worlds that might accompany their blueprints for India's new built landscape.[22] Their puzzle was not simply the form of buildings to design, but how particular design approaches and techniques might help to cultivate more desirable relationships between people, material life, and the urban environment.

In some ways, the architects I describe in this book—"green," or environmental architects—engaged in a practice of hope: hope that the urban future could be ecologically and socially reformed, and hope that their profession would position them in a way to enact that reform. At its surface, their endeavors might evoke ideas like those suggested by David Harvey, for example, in his *Spaces of Hope*. In it, Harvey directly addresses the theoretical figure of the architect to invite us to consider the social worlds that could inhabit the spaces architects imagine, and to invigorate the suggestive possibility of the utopian landscapes of which those sociospatial worlds are a part.

In this theoretical guise, however, the architect is often ascribed some degree of agentive power; we look to the imaginative sphere it signals as the source of new shapes for human history itself. Echoing this Marxian sensibility may leave the reader inclined to regard the architect's connection to built, material forms as automatically powerful, not only for imagining new social worlds, but also for distinguishing the human social world from nonhuman nature.[23] Yet, as this work will show, the engaged social world of environmental architecture is always and automatically suspended in a web of socionatural power relations, bureaucratic structures, and historical legacies that not only shape the architect's agentive potential, but the very imaginary itself. The work among environmental architects that I recount in this book aims to show how a set of social agents simultaneously composed important new visions of a more desirable Mumbai, and experienced structural limits to their capacity to forge from those visions the city of the present.[24]

In a lived reality of resilient and unequal power relations, in a city in which the material development of the urban form proceeds according to far more powerful actors than most urban design professionals, what compelled Mumbai-based architects to seek environmental training? What motivated them to enroll in a degree program that required significant commitments of time, money, and intellectual energy, but returned only scarcely discernable leverage to make change?

To these central puzzles, this book offers insight into the contemporary power of conceptions of the future, showing how shared notions of temporality emerge as critical for understanding the reproduction of environmental actions in the present. While a wide range of theorists have issued calls to take temporality seriously, and authors like Appadurai have persuasively established the place of the future as a "cultural fact," it is only through sustained attention to the everyday life of ecology in practice—here, as environmental architecture training and work—that we can come to appreciate the multidimensional role of temporality as it animates social structure and social agency.[25,26,27] My aim in this ethnography is to better understand and compose an architectural figure, as well as a contextualized actor, who is firmly embedded in the social structures and power relations of the present, and yet compelled by a specific and powerful set of temporal sensibilities to expect, and react to, a dramatically different anticipated future.[28]

I thus attend to the cautiously confident, deeply aspirational politics that took shape among a group of environmental architects in a Mumbai on the cusp of a new urban development plan. As I will describe, theirs was a politics of expectation and possibility that sought to defy both the triumphant pronouncements of Mumbai as a "development" mission accomplished, *and* repeated declarations that enduring inequality and intensifying environmental vulnerability sealed for the city a chaotic urban fate. I focus on a social arena of relative, but always compromised, privilege in which actors are neither fully empowered elites nor fully dispossessed. Unsatisfied with Mumbai's political economy and its environment, the architects I describe organized their aspirations to change both according to an emergent moral logic—a moral ecology that, as I will show, relied on the inevitabilities of the environmental future to reposition their active potential and to remake urban socioecological life.

Separating social life in the city from the biophysical vitality of the environment has long been untenable, so such separations are inconsistent with the lived social reality of the environmental architects profiled here. Across social theory and studies of social and cultural change, rejecting the modern human/nature divide has opened myriad theoretical and conceptual approaches to nature, and has helped us rethink our understanding of social change. Invoking ideas like "species being,"[29] "more than human geographies,"[30] or "multi-species ethnography,"[31] we are roundly challenged across disciplines and analytical postures to reconsider the intersectional arenas previously designated as humans here, in the city, and nature beyond—there, in the hinterlands. Marx's classical line between the bees and the architects no longer holds solid sway, bringing nature "back in" to political economic analytics and humans into old categories of nature. Indeed, as Dipesh Chakrabarty has argued, bringing nonhuman nature "back in" is no longer a discipline-based choice,[32] and conceptualizing agency exclusively in the human sphere is nearly impossible to sustain.[33]

At the same time, "urban nature" has gained new and globally circulating traction as a useful, and indeed often essential, conceptual component of the twenty-first century city.[34] Urban environments and their futures, concerns often referenced in shorthand through terms like sustainability and resilience,[35] are undeniably central features of cities as they unfold. Urban professionals who mediate this domain thus emerge as particularly needed, desired, and powerful actors on a stage of expertise emboldened by its claim to ensure and safeguard sustainability.

Or do they? The puzzle at the center of this book is one of *coupled* environmental and political transformation. Our lens is a collective of architects, brought together by a shared experience of formal training, and rendered a resilient, self-identified community of activist-professional-practitioners in the aftermath of that experience. They speak in this book of their passion for a practice that will eventually align social and environmental vitality, and they assemble key concepts in interaction and fellowship with one another and with the author, the anthropologist. The sphere of praxis they share merges a particular version of urban ecosystem ecology knowledge with design techniques, creating a science-design driven, shared point of reference that they repeatedly indexed as "good design." The book's later focus on the realm of post-training, lived professional practice allows us to trace a conceptual and experiential bridge between the work of the imaginary and the work of politics. Across the book, I draw from ethnographic data and analysis to better understand the relative power of architects as social actors who seek to integrate training and practice. What emerges, I will show, is a specific and important form of green expertise, but one that remains a vocation in waiting.

Despite the formidable social structures that condition their capacity to act in the present, I will describe how architects were nevertheless key agents of urban socioecological transformation in a city more often noted for its seemingly intractable *un*sustainability than for its demonstration that a different, more ecologically vital urban world is possible. Crucially, they remain agents in waiting: the configuration of bureaucratic power, urban development, and capital that composes Mumbai's political landscape ultimately suspends "good design" in a future still to come.[36]

GREENING THE URBAN REVOLUTION

Green architecture and design are expansive, conceptually and in practice. The terms invoke other equally broad concepts, including urban ecology, sustainability,[37] and urban nature. Like many malleable and oft-employed terms—globalization, modernity, and culture among them—green architecture and environmental design must be anchored to lived social life if we are to discern their form and meaning. In this study, that understanding is derived from the training and social world I encountered at The Rachana Sansad Institute for Environmental Architecture in Mumbai.

For analytical grounding, I employ an "ecologies of urbanism" approach, drawing from previous theoretical work with K. Sivaramakrishnan and the insights of the many colleagues who have been a part of our *Ecologies of Urbanism in Asia* projects.[38] This approach builds on formative thinking across several arenas of scholarship to propose studies of urban nature-making that foreground place and context. Rather than assuming a singular, universal ecology, and thus a unified experience of urban nature, our intention is to identify the multiple forms of nature—in biophysical, cultural, and political terms—that have discernable impact on power relations and human social action. In *Ecologies of Urbanism in India,* we wrote:

> Identifying and understanding these multiple forms is central to the analytic. Some hinge on human social processes, and some on non-human and/or biophysical ones. Each intersection may involve competing worldviews, aspirations, imaginaries, and assessments of the stakes of urban environmental change. Social efforts to ensure, create, or imagine ecological stability that characterize these intersections are often infused with ideas of political, social, or cultural improvement, revival, or restoration. To promote particular urban ecological futures, then, may also involve the reproduction or contestation of cultural ideas of belonging to certain social groups, territories (including the city, the nation-state, the region, and the realm called the 'global'), or, indeed, nature itself.[39]

At the same time, an analytical stance that is exclusively social is only partial, and quite unhelpful for the reasons discussed above. The ecologies of urbanism approach therefore demands attention to the underlying biophysical conditions and natural histories of a place, and it requires a multi-scalar perspective that varies its analytical parameters according to the social and/or biophysical processes under consideration. The result, as we write, may be for example that "the appropriate boundaries of 'the city' are not automatically known from municipal borders or demographic concentrations. Likewise, nation-state borders (may) not determine where and how a study begins and ends."[40] Our focus is thus on processes, and the imperative of tracing the scales and boundaries that the processes themselves compose. In this sense, the very connections that allowed my own ethnographic work to move from Kathmandu to Mumbai, discussed in the preface, extend from the idea of ecologies of urbanism.

While the analytic has proven generative in our efforts to understand urban environmental change in Asia, we recognize the enduring centrality of the biophysical sciences, which usually lay claim to the term "ecology" in its singular form. The biophysical sciences may offer only one in a constellation of competing and meaningful understandings of urban nature, and while each may enjoy a privileged or empowered social position at different moments, there is no question that regarding scientific ways of knowing biogeochemical processes and systems as unimportant leaves us with little capacity to understand socioenvironmental change. In this study, then, I pay particular attention to the specific concepts, methods and imperatives from ecosystem ecology that the architects assembled in order to compose a scientific basis for "good design." I ask what kind of ecological science the environmental architects learned. How did they employ that knowledge in their environmental design approaches?

This question is motivated, in part, by the many ways that environmental scholars across disciplines have sought to more fully integrate ecosystem science and social studies of the environment. A subfield of ecosystem ecology, the biophysical science of urban ecosystem ecology tends to follow theoretical and methodological innovations in ecology that include chaos theory, disturbance ecology, patch dynamics, and efforts to understand spatial heterogeneity. Urban ecology is not a science of fully fixed successional patterns, homeostasis, human "disturbance," and wholly predictive modeling that social analysts have historically, at times with significant consequences, assumed.[41]

In North America, two urban research sites among the US National Science Foundation's Long Term Ecosystem Research (LTER) initiatives have been particularly generative of urban ecology research findings and analyses.[42] These centers have long forged new ground in the scientific theory of urban ecosystems, and they have made significant contributions to the analytical tools available to scientists, social researchers, and design practitioners. An exemplary recent volume that captures some of the interdisciplinary accomplishments of this work, and its innovative models for urban ecosystems, is Pickett, Cadenasso, and McGrath's *Resilience in Ecology and Urban Design: Linking Theory and Practice for Sustainable Cities*,[43] but the wealth of particular and integrative studies produced in the Phoenix and Baltimore LTER's, as well as the many other ecosystem-science grounded urban ecology research consortia in North America and beyond, is vast indeed.[44] For the purpose of this project, I wish to note the longstanding efforts among ecosystem scientists to understand human social dynamics, and to meaningfully include them in their conceptual and research models.[45] Attempts to bridge natural scientific understandings of the way nonhuman nature works and understandings of how human societies work are neither new nor exclusive to urban ecology,[46] yet fully understanding how social and biophysical structures, functions, and agents mutually produce one another remains a complex and robust challenge.

What is striking to scholars in the environmental social sciences and humanities is the extent to which the science of urban ecosystem ecology has made it imperative to integrate human communities and human action into conceptual and practical models, not as automatic "disturbances," but as "natural" components.[47] Similarly, a notable aspect of many so-called green or environmental design interventions is the aspiration to integrate a sophisticated understanding of biogeochemical cycles, energy flows, and other landscape considerations into architectural thinking and decision-making.[48] Both worldwide and in specific locales, various sets of largely standardized metrics have emerged for assessing the degree to which individual buildings are attributed more formalized and quantifiable "green" status (e.g., LEED or BREEAM standards), but in an epistemological and practical sense, as will be explored in this book, environmental architecture transcends mere building codes and metrics. It encompasses aspirations for individual buildings, but also aspirations for transforming entire urban ecosystems in coupled social and biophysical terms.[49]

In the social sciences, urban ecology signals a vast, multidisciplinary body of work that might be clustered into many subgroups, just a few of which I detail here. In current anthropological work, particularly that which builds from anthropological strains of political ecology, efforts to theorize and analyze contemporary urban nature tend to follow longstanding theoretical discussions of "nature-cultures" and "socionature."[50,51] Rather than enumerate an exhaustive list, it is useful to notice here two nodes of convergence between urban and environmental scholarly praxis that have generated new understandings of the dynamics that coproduce social and natural change.[52]

The first node tends to locate its theoretical anchors in the Lefebvrian assertion that, by tracing the capitalist processes that knit together city and countryside, we are poised to recognize a "completely urban" world.[53] From this vantage point, urban political ecology might be characterized by its primary attention to the multi-scaled conceptual and material systems that organize the flow of capital, labor, information, and power. These systems include cities, but are by no means confined to, or defined by them. Geographers have been particularly prolific in generating such mappings, while anthropologists and other ethnographers have demonstrated the historical and sociocultural particularities of larger scale processes when they are enacted in specific places.[54]

A second cluster of contemporary social scholarship asks how the social analyses of the environment that developed in non-urban contexts might shed new light on our understanding of socionatural life in cities. Here, "urban" tends to signal cities and city life. While Lefebvre's broad urban processes are acknowledged, they do not automatically configure the field of inquiry. Field sites in this second group are usually located within or across specific cities or city neighborhoods, allowing researchers to explore how various forms of social asymmetry may be reproduced or reconfigured in the practice of urban environmental politics and management. By drawing from its legacy in environmental anthropology, this form of socioenvironmental inquiry affirms the fallacy of a clear rural-urban divide, but nevertheless takes the sociocultural and nonhuman natural life in dense human settlements to be distinctive from its non-city counterparts in significant ways.[55]

Both strains of scholarship emerged in response to three somewhat distinctive scholarly conversations in the social sciences, each a quest to rethink modern urban/rural and nature/culture binaries. One involved formally problematizing western analytical conceptualizations of ideal *nature* as located outside the city, and wholly separable from human culture; a second grappled with turn of the century globalism and economic globalization.[56,57] A third group, galvanized primarily through work in geography, proposed analytics for studying *the urban* in a way that emphasized large-scale, interconnected nodes of power, and the material and social flows between them. Here the importance of movement and the notion of a simultaneously human and nonhuman "urban metabolism" formed an influential theoretical basis for specific research approaches.[58]

Among political ecologists in anthropology and sociology, Amita Baviskar's proposal of a cultural politics approach to natural resources, and urban applications of theoretical debates about attributions of agency to nature—such as those posed by Timothy Mitchell, Anna Tsing, and many others—challenged social analysts of all disciplines to confront the untenable essence of fixed nature/culture dualities.[59,60] Among geographers, Castree and Braun's *Social Nature* laid useful groundwork for writing, as Braun encouraged elsewhere "a more than human urban geography."[61] This, combined with sensitivities to the political dynamics of scientific knowledge and knowledge production—and particularly to "systems" thinking—set the stage for recent ethnographies of urban nature and urban sociality that defy easy disciplinary classification.[62,63] Recent work by Timothy Choy exemplifies this new direction.[64]

The historian Dipesh Chakrabarty's previously mentioned, provocative call to rethink how scholars do their research in the Anthropocene Era brought the environment—its past, present, and possible futures—into sharp theoretical focus across the social sciences and humanities.[65] A call for disciplinary scholars to reconsider the place of nonhuman nature and biophysical processes in all manner of inquiry, this work underlined the impossibility of responsible consideration of nature without social life, and vice versa. Studies of urban ecology, to this mode of thinking, automatically demand contextualized, ethnographic approaches to urban social and biophysical change. In outlining our analytical approach, K. Sivaramakrishnan and I contend that these must remain interconnected with, and anchored to, historically produced social structures *and* imagined socionatural futures.[66]

But what is the relationship between analytical and theoretical approaches to urban ecology and an actual, lived life of environmental architecture? In this book, my dual focus on environmental architectural pedagogy and practice assembles an inquiry into the ways that urban ecology's prescriptives took social and material shape. I consider what conventional architects studied, and the extent to which they were able to do what they sought to do from their specific professional and historical positions in Mumbai. My goal was to observe whether and how ideas of what can and should be, according to the dictates of "good design," gave way to actual built forms. It is, after all, together that these comprise the material form of cities—precisely those things meant to be resilient and enduring in the biophysical and social fabric of a city.

A rich literature underlines the utility of distinguishing between architecture and design, as the historical, spatial, and political assumptions signaled by both are complex.[67] In its very basic sense, architecture usually points to the making of individual structures, while design and its close associate, planning, signal the broader, specific, and desired interconnections between them. Keeping in mind their very distinctive, and quite consequential, histories, the reader will notice that in this work, I tend to use these terms interchangeably. This simply reflects their usage by the environmental architects with whom I worked and learned at RSIEA; but it should not be read as a dismissal of the very important theoretical insights that scholars of architectural history, postcolonial urbanism, and nationalist politics advance when they disentangle these processes carefully.[68]

In this work, the reader will notice the ubiquity of the term "good design," which RSIEA architects employed to mark an ecological rubric for how a built form should be conceptualized, sited, and interlinked with underlying water, energy, nutrient, and waste systems. Such contextualization gave singular built forms an assumed embeddedness in environmental processes at a variety of scales, including site-specific questions about orientation to light or shade, as well as questions of placement in a watershed or a mosaic of land use and land cover patches. In this sense, environmental architecture and "good design" were two expressions of a simultaneously singular and scaled undertaking, and green expertise was distinguished by the commitment to engage complexity through thought across multiple scales.

Thus the reader will also notice that the phrases "green architecture" and "green design" are also used interchangeably in this book; here again, I do not wish to imply an analytical conflation of architecture and design. I follow instead the terms that RSIEA architects used to index the wide array of technologies, materials, and conceptual approaches to building practices that they believed would improve the overall sustainability profile of a building and wider site, a city, and onward to an interconnected city-countryside continuum of urban landscapes.

Conceived in this way, we might consider the assumption of automatic embeddedness in broader environmental systems as a sustaining logic for linking "good design" to a transformative movement. That movement, in both global form and place-based expression, often explicitly aims to change core concepts, forms, and practices of ecology in and of cities.

As an ethnographer, I draw guidance in this book from the lived experience of contextualized, everyday life in a specific historical moment, and in the social processes through which people consciously described and experienced city space in the making. My hope is to explore the coupled social and biophysical choreography, however suspended it ultimately remained in the realm of aspiration, that brought urban ecosystem ecology into harmony with architecture, and ultimately with urban social forms. In doing so, I notice and mark points of friction, meaning-making, and experienced limits.

It is with care that I chose architects as the focal community for this work. Despite the compelling place of both the theoretical figure and the social agent, cultural anthropologists usually defer to the expertise of archaeology and scholarship in

material culture to address the relationship between basic units of social organization and architectural forms, artifacts, and practices. Yet we have sometimes, intentionally or unintentionally, embraced rather rigid associations between built forms
and social forms. In fact, the very endurance of architectural forms has sometimes
led to fixed and determinative approaches to social explanation.[69] In this study,
imagining and making built forms are key social practices through which architects consciously and intentionally bridged their understanding of ecosystem ecology and their intentions for a more "sustainable" social reality in Mumbai.

Just as was true in earlier reference to the science-social science interface, the
idea that material form and social life are interconnected is neither new nor novel.
Among a host of examples in architecture and design history, the Green Cities
Movement and related experiments in environmental design were in part intended
to promote social vitality, and sometimes even social rehabilitation, by creating
ecologically contextualized built forms.[70] Likewise in this project, contemporary
green architects often expressed intentions to promote or enable revitalized social
configurations through their material designs. Unlike much modernist architecture, however, those intentions were grounded in an environmental restoration
agenda animated by twenty-first century concerns over environmental change and
the enduring postcolonial effects of deep socioeconomic disparities. In this context, the best designs would be responsive to biophysical uncertainties and sociopolitical imperatives, as well as to the conceptual values bundled in environmental
architecture training as "good design," as I will show.

Focusing on the interface of material forms and social relations can sometimes seem to reduce urban and political change to technical questions, but my
intention here is to do the opposite. The chapters to follow demonstrate the many
ways that socially meaningful aspects of green design were in fact far more than
technical, so much so that the most advanced technologies and materials often
assumed a background position in pedagogical and praxis-based designations of
"good design." In the foreground stood a more comprehensive moral ecology that
enfolded core ideas about what was right and necessary for the good of society and
the environment.

To appreciate this moral ecology fully depends on a careful treatment of the
ways that architects cultivated and operationalized the specific hybrid knowledge
form[71] they derived together in the context of training. That hybrid knowledge,
which was used to characterize environmental architecture as an "integrated"
subject, fused selected aspects of ecosystem ecology with equally selective design
technologies and social objectives. It was that same hybrid knowledge form that
distinguished the environmental architect from the architect, and in turn the
green expert from the urban professional.[72] Green expertise, however deferred
in its actual practice, nurtured and reinforced a shared hope that Mumbai—and
indeed, cities around the world—could be remade, and indeed could survive, in an
increasingly uncertain environmental and social future.

A GLOBAL ENVIRONMENT IN AN URBAN CONTEXT

Complex processes like urbanization, green design, and city life are often discussed as undifferentiated categories. Regarding them as universals, however, risks losing sight of a point advanced by Taylor and Buttel over two decades ago: that there are critical limitations to concepts, metrics, and simulations of environmental change conceived at the global scale when we seek to understand how *actual* change takes place in lived social life.[73] Context, they argued, exerts profound, if highly differentiated, influence on how and when eco-social processes shift.

Across global discursive and policy discourses, green design circulates with prominence and purchase. It is often invoked to provide alternative trajectories for housing and infrastructure development, energy regimes, and resilience planning. Yet our understanding of the social and political dynamics of green design in specific contexts is curiously limited. At best, we have only a preliminary understanding of the cultural and historical narratives from which green design derives its place-specific legitimacy, force, and moral authority. This book aims to more clearly define the contextualized social, political, and cultural processes through which green design knowledge circulates, transforms, and is operationalized— even if only in aspiration. By attending to these factors, we are better positioned to understand the structural barriers that prevent city-scale ecological change, and to calibrate initiatives to change the actual barriers they encounter. At the same time, we stand to gain a more sophisticated appreciation of the importance of sustained aspiration among those who stand trained and poised to implement specific kinds of ecological practices.

The potential impact of green design technologies and practices is obviously far-reaching. It is imperative, then, that we understand the particular fusions of scientific and social scientific knowledge that constitute their basis, and the moral ecologies, temporal sensibilities, and ecologies in practice that characterize their contextual variability. Clearly new relationships to the environment are being forged through the practice of green design; so, too, are new social relationships and wholly new political ecologies of cities and non-city spaces.

Toward that end, over fourteen cumulative months between 2007 and 2012 (including a period of eight continuous months in 2012) I used mixed social research methods to understand the social life of green architecture in Mumbai. I employed participant observation at RSIEA, in its Master of Environmental Architecture classes, and on the educational field trips that are part of the Institute's curriculum. I undertook additional participant observation among Institute faculty members, and among students who had completed the program and gone on to work in Mumbai as certified environmental architects.

To trace students' post-program experiences, I administered surveys to all RSIEA students, then-present and past. Of one hundred and five total graduates at the time of the research, I was able to survey ninety-six. Additional archival materials provided information on the historical context for environmental design in Mumbai, as well as its contemporary life as a pedagogical undertaking and a mode of professional practice. Finally, through a separate interview protocol, I administered twenty-seven semi-structured interviews and focus groups with students, and seven interviews among other relevant practitioners. Eight additional interviews were conducted with active RSIEA faculty members.

As a center for dynamic design in India, Mumbai hosts many environmental architecture training programs and professional groups. Several are better known, and by far more elite, than Rachana Sansad. While I learned a great deal about other programs, and met many people active in other arenas of environmental architecture practice and pedagogy, I focused my fieldwork specifically on RSIEA. This allowed me to concentrate my inquiry on a specific subset of Mumbai's architects who are committed to environmental design; it also enabled a richer contextual sense of the camaraderie that developed between these specific professionals who, once they graduated, continued to draw from their common experience of training and practice. A focus on RSIEA also allowed me to more fully experience, and therefore better understand, the academic curriculum through which ecosystem ecology was conveyed as, and then transformed into, design practice.

It is important to note that, although today RSIEA stands among many similar postgraduate programs in Mumbai, the Institute for Environmental Architecture was the first and only program of its kind in India when it was inaugurated in 2002. It was Roshni Udyavar Yehuda, the Program Head throughout this study, who authored the pioneering curriculum, and who has since watched it flourish alongside many others in the city.

Caste and class dimensions, ethnic and religious diversity, geographic mobility, and linguistic capacity also distinguished the group of architects I profile in this book. In contrast to more elite architecture programs in Mumbai, RSIEA is not generally considered the pinnacle of architectural study, either internationally or in India. It boasts successful graduates, but it is not generally associated with prestigious international firms or elite national and global connections. Indeed, those students from privileged and wealthier backgrounds, or with more international educational experience, tend to enroll and participate in other programs in Mumbai's vast architecture and design community. Focusing on RSIEA offers particular insight into the experience of Greater Mumbai-based, middle and lower-middle class professionals who were likely to remain in Greater Mumbai and Maharashtra after their training was complete. I use "middle and lower-middle class professionals" as a heuristic rather than a clear and constant category here, however, since the architects I learned among were nevertheless quite removed from the more glamorous and often transnational world of architectural elitism. Most came from Greater Mumbai, and remained in Maharashtra as practitioners after they completed their master's degree. They assembled from extremely diverse caste, religious, linguistic, and geographic backgrounds, but all had achieved fluency in English sufficient to train and earn their degree in that language. All had the capacity and willingness to enroll in a highly diverse, cosmopolitan setting for this training. Many took out formal loans to cover the costs associated with RSIEA, so they were also sufficiently socially positioned to gain access to formalized structures of credit and debt. In general terms, we may think of the architecture students at RSIEA as a set of bourgeois middle class professionals. Although not from the furthest class and environmental margins of the city— as was the case in my previous studies in Kathmandu—students at RSIEA each espoused specific caste, class, gender, and other conditions that influenced, in part, how and where their aspirations to make fundamental social and environmental change would be realized, and how and where they would face obstacles. As social actors, RSIEA architects were beholden to myriad established relations of power, constraints to mobility, and, in this specific instance, education or other forms of life-structuring debt. Despite their differences, as I will show, they cultivated what I have called elsewhere an environmental affinity group that allows us to consider them together, and to examine the moral ecological logics they assembled as they shared a commitment to "good design."[74]

. . .

The ethnography and analysis to follow are shaped by a set of core research questions. Each derives from a central and enduring interest in the ways that concepts of nature transform, and how that transformation relates to social and political life. I ask, how is the environment made, and made meaningful, in urban settings?

To ask how the environment is *made* is to highlight how the pedagogy of environmental design—the selection of historical narratives, notions of social duty and responsibility, aspects of ecosystem ecology that were conveyed and taught, and modes of transforming them into essential design techniques and skills—actually constituted a built form bridge between "more than human" nature and more-than-technical environmental design.[75] Understanding the form and content of that bridge is essential if we are to understand the origins of formal diagnoses of urban environmental problems.

Likewise, to ask how the environment is made meaningful is to recognize that human social action is always predicated on the making of shared values and meaning systems. Values and meanings are in turn often associated with particular visions of, and fears about, the present and likely future. Arjun Appadurai has noted that green design has a specific discursive quality that tends to bring the future into the present; this book explores how that consciousness of present and future made processes of identity formation (the *we* of the collective), narrations of history (in this case, defining *Indian* green architecture), and challenging or claiming the place of Mumbai on a map of cities that achieve prominence on the global stage, were all important avenues for attributing meaning to environmental architecture and the work of the architect.[76]

We have already seen how the environmental architect can be constructed as the vanguard of a social movement, a special subforce in service of changemaking in Mumbai. It is important to ask further, how did that vanguard define meaningful urban nature, and how did it seek to enact it through material practices? What moral ecologies and temporal sensibilities compelled them? For scholars interested in urban environmental change, there is perhaps no more fruitful place for answering these questions than among the architects themselves as they learned, debated, and sought to practice green design.

I assume in this work that architecture is a field of cultural production in Bourdieu's sense; there is no single and stable voice of all architects, but there are nevertheless important agents and institutions. In the present case, this would include, among many others, the All India Council for Technical Education, which was reformed in 2009, the Council of Architecture, the RSIEA-accreditor Yashwantrao Chavan Maharashtra Open University, and even the Archaeological Survey of India.[77,78] The actions of these institutions simultaneously produce specific cultural goods (in this case often by guiding and certifying a curricular structure) and those interested in, and positioned to, consume them. Many other relevant agents and institutions also exist, and exercise influence, at a variety of scales, from the global (such as the internationally recognized BREEAM metrics or LEED standards) to the regional (such as GRIHA).[79,80,81] Although regularly contested, these institutions and agents were also regularly invoked as sources of legitimacy, and so produced and reproduced the templates for specific kinds of built forms, and those who sought to consume the attributes of those forms.

In the next chapter, I set out to understand the social life and practice of environmental architecture at Rachana Sansad Insitute for Environmental Architecture, first through the genesis narratives and curricular goals espoused by founding faculty members in the Institute, and then by noting the profound—almost unbelievable—sense of optimism and possibility that formed their basis. The study continues to explore how "good design," a quality foundational to the RSIEA curricular mission and yet very diffuse, was conceived and conveyed. These opening sections trace social and pedagogical life at RSIEA through an entire curricular cycle, following students on mandatory field trips, accompanying them in the classroom and in project work, and describing the gradual formation of alliances and affinities that would extend far beyond the two-year master's program. They show how basic conservation biology and systems science principles formed the scientific basis for green architectural pedagogy, while specific narrations of history, vernacular tradition, and climatic specificity attached the science to locally grounded techniques and practices.

Chapter 3 begins by describing the opening session of Rachana Sansad's 2012 Environmental Architecture course, which featured a collective screening of the American climate change film *An Inconvenient Truth*. It examines the distinctive blend of globally and locally circulating technical material that was hybridized to produce and convey a Mumbai- and India-specific concept of good design.

By this point, the reader will find the broader context of the city noticeably absent, and so two chapters follow that refocus our attention on the broader urban and temporal contexts in which RSIEA architects trained. In these chapters live debates about greening the city and contrasting public exhibitions calling for new urban plans and designs sharpen our view of the public spectacles that characterized contests over Mumbai's urban form and future.

These chapters also reposition our attention onto the concept of open space, an omnipresent, quite popular, and yet deeply problematic prescriptive in this period. Departing for a moment from my direct attention to RSIEA in order to follow one of its faculty members to an important urban forest patch, I trace the politics of exclusion embedded in aspirations and practices of good design.

Fortified with an ethnographic snapshot of RSIEA and a sense of the urban context that enfolded it, Chapter 6 then follows RSIEA students through a key aspect of RSIEA pedagogy: their ventures outside the city. Centrally important field trips, like the BCIL journey I described in the opening pages of the book, played an important role in synthesizing the idea of a distinctly Indian history, quality, and imperative for good design. This section considers some of the sites, their complex histories and symbolics, and the ways that encounters there were structured and limited by environmental pedagogy. This chapter shows that regardless of the presumptive power of sustainability to render such places instantly neutral, producing green expertise was a history-claiming endeavor that depended on a culturally grounded spatial geography to enchant and make meaningful a more globally legible choreography of technical training and skills.

From the training journeys far afield, Chapter 7 returns to Mumbai to trace how good design aspirations fared in domains of practice. The reader follows RSIEA graduates, and students on the cusp of graduation, as they seek to turn environmental architecture into ecology in practice. Interviews with a range of graduates bring the core tensions between the structural forces of urbanization and the aspirations of RSIEA's green architects to life as we encounter the complex of power relations and bureaucratic structures that modify grand plans. We also trace the resilience of the moral ecologies and temporal sensibilities that maintain green design as a tool of future salvage and a marker of identity. While the structural elements of economic growth, new building construction, and socioeconomic development in Mumbai constitute several books in their own right, this chapter sketches their formative contours by noting how they converge with green architects' agentive efforts.

The contextualized, structurally conditioned figure of the architect returns in the book's conclusion; so too does the importance of moral ecologies and temporal sensibilities for making dormant aspirations into resilient forces shaping social

life, social change, and ecologies of urbanism in Mumbai. Indeed, whether studying social change or ecological process or both, our tools of analysis are in a flux
that mirrors the lived realities profiled, and suspended in layers of context and
power, in this book.

The Integrated Subject

"At the Institute of Environmental Architecture, we believe that the architect's primary role is stewardship of the land & environment. It is in establishing this intrinsic co-relation between human beings and the rest of the natural world that the architect's creative abilities are realized."

—VISION STATEMENT, RACHANA SANSAD INSTITUTE FOR
ENVIRONMENTAL ARCHITECTURE

As I prepared my first invited talk at Rachana Sansad Institute for Environmental Architecture, my thoughts fixed on the presence of a mutual question between me, in the role of an invited speaker assumed to be worthy of at least an hour of focused attention, and the audience, a group of professional architects assembled as students of environmental architecture. I'd never trained in architecture, and I'd never met the students or faculty member prior to receiving the lecture invitation. Still, as I recounted in the preface, the social and ecological processes foregrounded by culturally charged debates over urban housing, informal shelter, and built space in general had led me to Mumbai, and had, quite unexpectedly, brought me among them as a guest lecturer.

In a fit of nervousness that my lack of an architecture background might prevent me from sufficiently connecting with the students, I found myself settling on the reassurance of a shared, and always contextually informed mutual question: "What does urban ecology *mean?*" I had used this question to help me organize my analysis of the very case study my lecture would present, and I realized it was also likely shared with this new and unfamiliar audience. For the architecture students, the meaning, or rather meanings, of urban ecology would form the basis for my urban ecology perspective on the Kathmandu case. For me, it was a matter of opening that same question afresh. The architects enrolled in an environmental architecture program presumably sought a kind of training that would enable certain new perspectives, insights, and forms of knowledge. In sharing my work, I thought, perhaps I'd also get a glimpse of the kind of practice that constituted an urban ecology approach in architecture in contemporary Mumbai.

Beyond the lecture hall, of course, public discourse in South Mumbai buzzed with discussions of the city's new development plan. For this short but animated period, a wide range of the city's publics found it a bit less preposterous to imagine new, more ecological built forms animating the future urban landscape, supplemented, perhaps even generously, by new open spaces. I held that in mind as I made my way to Prabha Devi, climbed the open, airy staircase of Rachana Sansad's main building, and met a group of RSIEA students and faculty for the first time.

At the close of the presentation I invited questions, and several students responded by offering comparative reflections on similar cases in India. A range of design and spatial concerns that were not part of my usual analytical impulses emerged in their detailed critique, giving me a first experiential glimpse of the conceptual and technical dimensions of RSIEA's pedagogy of environmental design. My place as an anthropologist of environmental architecture was thus quite clearly rendered: as I learned from my interlocutors, and was consulted as a sort of "green expert" whilst discussing the ways my expertise fell short, I would also interact with them intellectually and personally. Each encounter made the context even as I sought to document and understand that same context, and I was soon engaged in an ethnography of the training and practice of environmental architecture at RSIEA.

Immediately afterward, I scribbled notes and questions, puzzling over the unfamiliar references and design analytics this group of student professionals brought to bear on my lecture. Specific epistemologies of environmental and social change, and the suite of techniques their profession might apply to the case, seemed to ground their shared expectations of what urban ecology meant in the Kathmandu I'd just discussed. Here, I thought, is perhaps an emergent praxis, an ethnographic understanding of which would depend on attending to the ways that experiencing RSIEA training would codify, activate, and enable a process of translation between a domain generally associated with ecology, and the agentive practice of environmental design.

Soon after that first lecture, this study unfolded. I first focused intensely on experiencing and understanding the pedagogical model employed at RSIEA. By following an entire curricular cycle among the students, I noted the program's form and content, its geographic reach as charted in course study tours that extended far beyond Mumbai, its programmatic flexibility, and its periodic moments of fixity. Completely new to architecture in discipline and practice, I tried to grasp key substantive components, metrics, and the sources (alternately global, regional, and grounded in smaller scale places) from which they derived when invoked in their "environmental" guise. As I attended classes (hurriedly taking notes on topics like thermal comfort and design or the environmental efficiency attributes of India's regional vernacular forms); traveled to places like Chennai, Bangalore, Koorg, and Auroville to study notable examples of regional environmental architecture; and walked the field site for a capstone design assignment in Pali, I was simultaneously a student, a professor, and a researcher. My life and livelihood stood apart from the real life stakes of mastering the training and excelling in its aftermath, yet I felt ever more invested, both intellectually and personally, in that same training and its outcomes.

This decidedly unusual field position is not without unresolved complexities. I am not technically trained in architecture or architecture education, and so this book does not pretend to analyze the work of RSIEA architects according to a specific architectural or educational theory. Nor is it an operational assessment intended to gauge the program's relative success or failure. Furthermore, to design the study using RSIEA as its epicenter meant that the singular language of the Institute and much of the profession—English—left many undoubtedly important dimensions of the place, situations, and layers of contests within them either obscured or entirely omitted. The scope of the study, and its potential to address specific questions, then, must be acknowledged as inevitably partial and incomplete. The reader who seeks nuanced analyses of the issues this book cannot address may find it lacking, yet I hope nevertheless convinced to enrich the observations and analysis herein through further attention and study.

. . .

In the bustling commercial and residential neighborhood of Prabha Devi, Rachana Sansad hosts a variety of undergraduate and graduate degree programs in several urban and design fields. It was founded in 1960 as an Academy for Architecture, and the school gradually introduced undergraduate and graduate programs in art, interior design, fashion and textiles, construction management, urban and regional planning, photography, music, and event management. Rachana Sansad is a school bustling with young professionals who have returned for advanced study or training, as well as students receiving their first professional degrees. It draws from across Greater Mumbai and Maharashtra, and administers all of its courses in English.

Among its graduate programs is the Rachana Sansad Institute of Environmental Architecture. Although an Academy of Architecture was founded within Rachana Sansad at the time of its inception, the Institute of Environmental Architecture was not established until 2002. From the first semester, which saw an enrollment of just two students, the Institute has grown to host forty students per year. Each undertakes a two-year, master's level degree program and earns a postgraduate degree in Environmental Architecture. A rolling roster of roughly thirty-five visiting instructors joins a core faculty of four—three trained as architects and one as an environmental scientist—to teach courses, lead study tours, conduct field project work, and evaluate student performance. The Institute maintains its official university affiliation with the Yashwantrao Chavan Maharashtra Open University in Nashik.

RSIEA's public vision statement, declared in printed literature and on its website, describes a conceptual mission in which architects regard their professional actions as automatic environmental disturbances, much in the way that early characterizations of ecology and nature regarded human activity in terms of perturbation and impact.[1] The statement reads, in part:

> When architects construct buildings, it has an impact on the environment. It affects the ecology of the place, disturbs the flora and fauna, changes the course of water bodies, pollutes the air and depletes finite natural resources. Is such destruction imminent and inevitable? We don't think so. At the Institute of Environmental Architecture, we believe that the architect's primary role is 'stewardship' of the land & environment. It is in establishing the intrinsic co-relation between human beings and the rest of the natural world that the architect's creative abilities are realized.[2]

At the time of this study, most of the students who enrolled in the RSIEA master's program did so while continuing to work in professional architectural firms and smaller practices. Few could afford to sacrifice their income to devote exclusive attention to the course work, and the schedule of classes is designed with this in mind. Nevertheless, the travel-intensive study tours that form an integral aspect of the program, and are optional, are historically heavily enrolled. Students may not be financially positioned to leave employment completely to undertake the course, but the level of personal and financial commitment was significant, a point to which I will return in a later chapter.[3]

Students come to RSIEA from varied ethnic and religious backgrounds, and faculty often remarked that they hoped that the religious and cultural diversity present in the classes would mirror that of the wider Mumbai population. The male-female ratio slightly favored women at the time of the fieldwork, but by 2017 women constituted 90% of new enrollees. When I asked faculty members why they thought women's numbers were rising so dramatically, answers tended toward noting that such statistics change from new class to new class of students. Conversations with female students, on the other hand, emphasized that continued graduate study in any field often allowed young women to delay getting married, even if only temporarily. University admission is officially described as per merit and government reservation policies, but at the time of the fieldwork I met very few students from the social categories officially considered "reserved." The academic profile of students tends to be strong, with most having achieved a position in the "first class" as undergraduates.[4]

At first glance, RSIEA's degree program in Environmental Architecture looks extraordinarily ambitious, perhaps so much so as to be quite unrealistic. A wide range of courses not only covers basic concepts in the ecosystem and environmental management sciences, but also an extensive array of environmental topics in design technologies and techniques, information systems for landscape analysis and mapping, law and policy, and the social sciences. A quantitative methods course is included in the required curriculum, as are group field projects and a fully independent final thesis. In addition, RSIEA regularly organizes and hosts public programming that explores contemporary questions in environmental design. The Institute's location in Mumbai allows it to draw from differently trained and positioned voices to address questions of urban sustainability, urban development, housing, green building techniques, and policy reform. While far from the most prestigious architecture graduate program in Mumbai or in India, the Institute is widely recognized as the first of its kind, and is in this sense a path-breaking pioneer.

Like its students, the core and visiting faculty are also practitioners in their relevant fields. Their demographic and cultural composition varies from year to year, but during my field work period the faculty's male-female ratio hovered in an almost even split. Instructors were of various ages, from early thirties to late sixties, and their cultural backgrounds included Gujarati, Marathi, Tulu, and Muslim. In addition to teaching together, several faculty members also practice professionally in the context of the Institute's Research and Design Cell, which regularly serves as a source of case studies used in RSIEA courses.[5]

. . .

How does one forge an environmental steward from a practicing architect? More precisely, and consistent with the Institute's stated mission, how does one assemble a pedagogical bridge between conventional architecture and its environmental alternative? As the first program to attempt to build that bridge in India, it seemed important to understand where RSIEA came from and the logic that brought its founders together.

Tracing the Institute's genesis narrative might begin with its founding scientist, now a senior core faculty member. A modest and observant man in his sixties, Dr. Ashok Joshi holds a master's degree in zoology and a Ph.D. in environmental science. He has never received formal training in architecture, yet he described the earliest germs of the RSIEA idea as the product of his conversations with architects about the absence of an integrative way of thinking about ecology and built space in Mumbai. Recounting his personal intellectual and professional trajectory, he described a specific kind of learning that he experienced in settings which brought multiple disciplinary approaches to environmental questions:

> When I was doing my master's degree, we had a scientific program that was overseen by the United States National Science Foundation . . . it involved students from different departments and . . . we took up the issue of pollution. They said that this was an integrated subject; it cannot be (addressed) by (just) one department. So it involved botanists, chemists, zoologists, and people from all different branches of the sciences. . . . (From) that I learned . . . the basics of environmental studies. And then for my Ph.D. I took up an environmental problem: the effects of pesticides and human wastes on fish . . . . From 1976–82 I was working as a scientist on two different projects dealing with ecological impacts. . . . Around that time, one of my friends, an architect, started discussing the environmental aspects of architecture. He would

FIGURE 2. Dr. Joshi delivers a lecture on rainwater harvesting in an RSIEA classroom. *Photo by the author.*

> ask me questions about plants, water treatment, or ventilation . . . and when I asked him why (he had not learned these things) in his architecture training—you know, at least some basic things about the environment—he said, "well, it's just not part of it." I understood very clearly then that there was a huge scope for architects to learn about the environment. So we combined his architecture knowledge and my ecology knowledge and we came to Rachana Sansad. We asked them if we could start a course, and we proposed it to the Maharashtra State Board of Technical Education. . . . Eventually we framed an entire (curriculum) and the Board gave us permission to give a diploma. . . . So for about three years we were giving this as a one-year course.[6]

Soon the one-year program became a two-year, master's degree-granting program. He continued:

> For three years we did that course, and after that the YCMOU (the accrediting university) expressed an interest. So we expanded the course and (curriculum) and got approval for a master's degree—two years. . . . Initially we had only two students, but when it became a master's degree, we had an intake of twenty.[7]

The program was first offered in 2002, marking a shift in formal architectural pedagogy in Mumbai. Prior to this program, there was no codified way to undertake formal architectural study in India that focused on how architectural approaches and practices intersected with the environment as, in Joshi's words, " . . . an integrated subject."

Since creating that integrated subject can risk undermining the explanatory power of the knowledge forms it combines, our conversation turned to the obvious challenge of providing adequate coverage of the vast intellectual terrain signaled by the idea of the environment as "an integrated subject." Joshi explained that the key was to assemble many voices of specialized expertise, and to build a curriculum that amplified one focused voice at a time. The functionality of the curriculum thus relied on identifying and hearing from those specialized perspectives by drawing from the extensive social and professional network each founding faculty member maintained. Joshi explained:

> We knew we would need scientists from many fields, and fortunately in Mumbai we have several. . . . We had good social contacts. We knew excellent, experienced people. And we employed several experienced people who are working in the field to teach different courses. And more fortunately, people were also interested in coming and sharing that knowledge. I actually can't say why, because monetarily it was not at all remunerative, but somehow—is it is out of luck?—for example, one Dr. Latoo, who is renowned in botany, came regularly to give lectures about plants. And Arbinash Kubal, whom you know is the expert of Maharashtra Nature Park, he . . . also came. It was like that. For geology and geography, the head of the department at Bombay University came. These are very (senior) people, very, very experienced. They came. Regularly. Like that we had more than fifteen resource persons who would come and teach specialized courses and lectures. . . . My responsibility was teaching the course that gave the basics of . . . ecology. . . . Once the ecology concepts are clear, only then can you apply them to architecture.[8]

The pedagogical strategy depended in part, then, on the place of its founders in that same social network of environmental specialists and practitioners, the "good social contacts" who could provide positioned expert voices to be aggregated through the curriculum itself. Guided by the RSIEA's specific Environmental Architecture curriculum, students would cultivate the skill of weaving together specific knowledge forms, and discern which were most critical for a given environmental design decision. In this way, ecology in practice in the form of environmental architecture involved mastering a strategy for assembling expert views deemed relevant, and distilling those views into architectural ideas, plans, and drawings.

But Joshi's comments pointed as well to the forces that compelled the members of his social-expert network to participate as contributors to this integrated learning endeavor. Clear and consistent financial gain could not explain their participation; for Joshi, it pivoted instead on shared devotion to a vision of integrated environmental architectural practices that could create positive ecological outcomes. Teaching was a primary way to promote and enact this vision.

The M. Arch. (Environmental Architecture) degree is today a two-year master's degree program that is conducted, as its brochure and website convey, in "full time face-to-face counseling mode over four semesters of 15–18 weeks each (with approximately twenty hours of contact sessions per week)."[9] Students who have earned a B. Arch. or its recognized equivalent are eligible to apply. The curriculum evolved slightly over the course of the project, with a revised course program differentiated as the "new syllabus." In both the previous and the newer curricular formats, students complete courses across a vast interdisciplinary landscape; they move between theory and practice, quantitative and qualitative modes of data measurement and assessment, and wide analytical techniques at various spatial and social scales. This remarkably broad course structure also covers specific laws, policies, and metrics that govern conventional and environmental architecture in India and worldwide. Even in its earliest form, the program is an ambitious curricular attempt to forge and teach environmental architecture as "an integrated subject."

If the vast content prevents students from developing rigorous theoretical or methodological depth in any one disciplinary arena, it nevertheless exposes them to a wide range of intersecting issues that fall under the general conceptual rubric of ecology. In a more contextualized way, it also traces a set of expert figures active in Indian environmental research and management: the same social-professional network from which Joshi drew to run the nascent curriculum remains a major source of visiting faculty members who staff courses each year. The integration, then, extends beyond the curricular content of a subject area called environmental architecture; it populates the integrated subject field with a set of practitioners to whom students not only attribute the status of an expert, but also with whom the student might be inclined to confer once graduated and practicing in the field. Joshi framed this in terms of simultaneously knowing the limits of the architect's capacities and cultivating the skill to discern the quality of knowledge derived from fields outside of architecture:

> In this program they get a total understanding and they learn their limitations. They learn that they cannot do everything on their own, so they learn how to get resources—from where can they get the right (information) to give you, say, an example of responsible wastewater treatment or good solid waste management. These are both specialized fields that we present—not in detail, but how to get the exact things you require for your project. They learn that they must evaluate whether the solutions specialists offer are correct or not, good or not. They must know how to test them. So our focus is on making them understand these things on a larger scale, to evaluate according to the principles of ecology and the environment.[10]

There was a tension, then, between imparting the importance of consultation among a network of appropriate experts, and cultivating the capacity to "evaluate whether the solutions specialists offer are correct or not . . . " As I moved through the RSIEA training experience, it became clear that we were simultaneously learning to forge a strategy for aggregating diverse knowledge forms, and to determine which knowledge forms mattered and to what degree. This evaluative authority derived not from a position of commensurate skills and training, but rather from the distinctive expert position of the environmental architect: a good environmental architect possessed and employed a capacity to think across the range of scales and biophysical processes that environmental architecture signaled. These might encompass biophysical and social details relevant to a building site, its broader biophysical and social context (such as a watershed or a bounded city), and a site's interconnection with processes that nested those scales into even wider contexts (such as an entire river system or a demographic migration pattern). Here, we might characterize the curricular mission as fostering the capacity to think in an "integrated" way across disciplinary and scaled perspectives, and to privilege this capacity over, for example, specializing in aquatic chemistry, urban sociology, or so-called conventional architecture. These green experts were first and foremost integrators; they were not architects who had mastered environmental science or social science or both. The sources and types of knowledge they would integrate, and the frameworks through which the quality of that knowledge was deemed acceptable were critical. At RSIEA, that framework was often shorthanded simply as "good design."

Learning to discern and undertake good design was a key intended outcome not only of the formal experience of the curriculum, but also of the social experience of collective learning and application. As faculty and students forged and experienced the shared conceptual space of good design, they came to share a collective, cultivated environmental subjectivity that valued and sought to elevate the Institute's specific approach to the built form.[11] As both an integrated—that is, interdisciplinary—subject and a collective social experience of learning good design sensibilities, then, the "integrated subject" produced an integrated subjectivity. Tracing and examining this subjectivity, and noting the environmental affinities through which it operated, is part of the work of the chapters to come, but let us mark here Joshi's characterization of the impulse to name, define, and employ "good design," and its attachment, via the RSIEA curriculum, to existing networks of expert knowledge.

Another important dimension of the RSIEA mission was the perhaps less tangible, but nevertheless central, notion of devotion or commitment. This was at times described to me in equally integrated terms, in the sense that teaching environmental architecture and being an environmental architect were part of a "totalizing" lifeway. Clear lines between vocation and job, personal and professional, or work and politics were always elusive. Faculty, and occasionally students, invoked their strong commitment to environmental architecture and its potential outcomes in order to explain, perhaps far better than Joshi's characterization of it as "luck," why such reputable scholars and practitioners would detach their work for RSIEA from direct and fair payment for their service. Such "commitment" was not exclusive to guest lecturing experts who might forego an honorarium; I regularly

FIGURE 3. Dr. Latoo talks with RSIEA students during a field study visit. *Photo by the author.*

noticed faculty members devoting considerable personal time to students, to course development, and to the Institute. Few voiced open concern for quantifying the hours spent or enumerating tasks performed, and the few I observed who did tended not to remain with the faculty.

An interconnected set of exclusions and self-selections are therefore also apparent here. Those who undertook the committed practice of teaching at RSIEA were often, though not always, already identified as experts within a specific socio-professional network, and some were also occasionally able to forgo payment (or accept a smaller amount than might otherwise be offered) for hours worked. While these were not universal attributes for every faculty member, they figured prominently, and were certainly central to the early life of the program.

· · ·

One morning in July 2012 I arrived at the Institute to find the core faculty filing out of a meeting room. Their faces were unusually somber; some bowed their heads. Something had clearly transpired to produce collective disappointment.

I learned a few hours later that a curricular evaluation session with representatives from the Council of Architecture Certification Board had returned some unwelcome news. In response to growing enrollments, the Institute sought to activate a revised curriculum that would facilitate certain forms of faculty and programmatic expansion. Yet the Council had rejected their proposal on at least two grounds.

Up to this point, mid-2012, work and excitement surrounding the Institute's expansion had been part of the everyday atmosphere among faculty. Countless hours, meetings, and discussions had brought the collective project of developing a new curriculum into the center of an already crowded workload, but people proceeded with confidence that a new course and scheduling sequence would enable more in-person contact between students and faculty, and more effective experiential learning. The new curricular design would place stronger emphasis on individual student projects, and culminate in an independent final thesis. New policies would make course attendance compulsory and admissions decisions wholly merit-based. Although faculty members voiced a commitment to keep tuition fees at a level competitive with like programs in Mumbai, it was clear that new teaching staff would have to be hired, and so a tuition increase was imminent.

Having witnessed doubled enrollments over only a few years, enacting these changes carried a certain urgency. But the impasse at the CoA meeting seemed to indicate that something had gone wrong; institutionalized structural requirements had somehow clashed with the program expansion proposal.

After everyone had settled back at their desks or rushed off to classrooms, I shuffled into Roshni Udyavar Yehuda's office. She explained that the CoA objections related to violations of some of the basic requirements for academic degree programs in any kind of architecture in India. The first objection was that current Institute faculty were not paid the Council of Architecture's regulation salary, and in order to hire new faculty, all compensation levels would have to be adjusted to those regulation levels. Even with proposed tuition increases, Yehuda explained to me, this was completely unfeasible. "We simply can't afford it," she said plainly.

The core issue was more than a matter of budgetary calculus, however. Yehuda lamented that reducing the value of teaching environmental architecture to its price in INR diminished what was an otherwise expansive endeavor "far beyond economics." She repeated that the RSIEA faculty had, since its inception, taught "out of devotion to the subject,"[12] emphasizing that the faculty had never joined because of the lure of the salary. Furthermore, she argued, "How can we say that someone who makes less in salary is less competent, or less valuable? By that score a volunteer, or someone who comes for the sheer passion of the teaching, is automatically not competent in the eyes of the CoA."[13] Her voice betrayed exasperation.

She continued: "It's the culture of this program that we teach here because we are committed to the subject. The salary is not a measure; it has never been the reason we teach this."[14] While for her this was a way of ensuring that those who taught at RSIEA were fully committed to the mission, it might have also prevented those without prior material security from considering it, regardless of their passion and commitment.

Yehuda's reaction to the Council's concern arose in large measure from a frustrating collision with the structural limits of what, up to that moment, had been a differently regulated collective space, one centered on a shared willingness to accept benefits other than money in exchange for environmental architecture teaching work. There existed no metric for capturing, conveying, and affirming this kind of value for the Council, leaving the faculty "devotion" so central to teaching at the Institute supplanted by a more powerful, if external, regulatory protocol. The failure of monetary metrics to capture the myriad forms of value represented by the environment is a longstanding theme in environmental studies; indeed, the problem of "commensurability" forms an important basis for a wide array of critiques in political ecological theory.[15] In this instance, the externalities were social. Yehuda lamented the absence of adequate metrics for seeing and valuing a collective social mission; while faculty members were officially laborers, they were socially fellow devotees. Of course, in the eyes of the CoA, such conditions simply constituted labor exploitation, and could not proceed.

A second Council objection underscored the incommensurability of the "integrated subject" and the CoA's measures of professorial fitness and curricular integrity. Regulations defined faculty eligibility strictly according to degree status, such that appointed faculty must hold degrees in architecture. Perhaps ironically, this disqualified nearly everyone on the faculty and left only three core faculty members. Despite accomplishment or expertise in their given disciplines, most professors could not, by this definition, remain eligible to teach at RSIEA. Again, the regulations presented a structural obstacle to a core principle of the integrated subject: "Our strength and uniqueness is the fact that this program is not only one discipline. We depend on that," Yehuda said. "Its entire future is threatened now." I asked naively if anyone had expected this challenge to the Institute's expansion proposal. The response was a troubled, almost blank stare. "No."[16]

The integrative improvisation that had marked the Institute's inception and so much of its history was possible at a different scale, with smaller enrollment numbers. Bustling enrollments, ever-growing demand, and the expansion of the professional field itself re-scaled the undertaking so as to render it more legible to institutions whose principles of value and indicators of merit clashed with what Yehuda called RSIEA's "culture." With legibility came new layers of scrutiny, potential sanction, and frictions between an idealized, integrated learning domain and the regulatory structures that extended from the political economic context beyond Rachana Sansad.

. . .

Many RSIEA faculty members brought to that moment a shared, decades-long history of creating and operationalizing their vocational curricular mission. Udyavar Yehuda's reference to "devotion" captured a prevalent characteristic of their interactive mode as colleagues and as teachers. Many were also constant collaborating practitioners and close personal friends. Their work together was often organized in professional terms, but also functioned between with the many textures of sociality: close friendships, shared projects, and the ubiquitous shared mission made it difficult to regard RSIEA faculty as simply a collection of environmental architecture teachers. Indeed, at times these social textures could be accurately framed as quietly political—a kind of professionalized, but tempered, environmental activism.

In addition to teaching, a subgroup of RSIEA faculty worked as practitioners under the auspices of the Institute's Research and Design Cell.[17] On a day spent walking the urban landscape together in the Matunga neighborhood of Mumbai, Yehuda remarked to me that between teaching together and working on projects together, the faculty was "practically like family."[18] In a separate meeting months before, she'd mentioned with pride that India's *Outlook* magazine had recently ranked Rachana Sansad fifth among architecture academies in India. What made RSIEA completely distinctive, she told me then, was its faculty and their extraordinary commitment to the subject and the mission. "We work together in every way," she said at that time; "We're like a functioning family."[19]

Many faculty members repeated strains of this sentiment; the shared mission reinforced the quality of the personal and professional relationships through which it was enacted. Their devotion to the integrated subject reflected an integrated subjectivity, albeit in a different register. At the same time, we rarely spoke of the few faculty members who joined the faculty during my research but decided to leave. That they existed reminds us that the interpersonal and professional affinity, and the standards of "devotion" and commitment were neither automatically desirable nor universally possible to meet.

· · ·

But how and when does one move from the more bounded category of a qualified teacher of some aspect of environmental architecture into the "totalizing" lifeworld of RSIEA's "integrated" subjectivity? When I asked the Program Head when she first began to sense that for her, environmental architecture would transcend a simple job, she traced her response to a single figure from her past. Decades earlier, Yehuda had worked with an organization headed by the Indian environmental activist Rashmi Mayur. Mayur was the founder of India's International Institute for a Sustainable Future, and had served as an advisory figure in the key UN Meetings that had shaped the international environmental policy agenda in the early nineteen-nineties. This was a time of new forms of environmental thinking and discourse at the international scale; emboldened by the formulation of sustainability espoused in the pivotal Bruntland Report, landmark meetings like the first Earth Summit in Rio de Janiero (1992) and the Habitat Summit in Istanbul (1996) carved new concepts for understanding the interface of environmental change and socioeconomic development. A wholly reworked agenda for global-scale issues including poverty alleviation, biodiversity preservation, and environmental conservation followed in their work. Mayer was present and active in international environmental policy circles in this moment, and Yehuda recalled her experience of working with him as deeply formative.

In a letter to me that accompanied a gift copy of *Survival at Stake,* the anthology of Mayur's work that she co-edited with a colleague in 2006, Yehuda described him as:

> . . . known to everyone—from the Prime Ministers of several countries to villagers. He was very popular, as he had worked on some major environmental movements, including the Bhopal Gas Tragedy, and he was responsible for the closing of some large polluting industries. It was on the invitation of Indira Gandhi some time in the early 1970s that he came back to India after completing his doctorate studies and joined as Director (of the organization).[20]

In her introduction to *Survival at Stake,* she called Mayur a "visionary" whose:

> . . . spirit lives on in the souls of thousands whom he inspired to tread his path. Popularly known in India as the "doomsday professor," Rashmi Mayur prophesied that if human beings continue on their present reckless path of mindless development, the earth's ecological systems would collapse and the human race will become extinct. (He wrote) "The consequences of the war that has been waged against this planet for the last two hundred years by human beings, may be that we may have no human inhabitants in the future." However, unlike many crusaders who relinquished hope and left the battlefield, and others who refuse to recognize the symptoms of a diseased planet, Rashmi loved it enough to see it with the eyes of truth. He was too optimistic to be biblical. "Nonetheless, we cannot be immobilized by the ugly reality. As long as we are alive, as long as we have vision and as long as we think of the future of the earth and our children, we must hope that sanity and wisdom will prevail."[21]

The essay continues to narrate Mayur's basic biography, and emphasizes in particular his place as a "world citizen" whose commitment to amplifying environmental causes had lasting effects in India.

That RSIEA's Head traced her own devotion to her work with a distinguished environmental activist in India and at the United Nations is consequential for understanding the form and mission of the Institute itself. The challenge to which RSIEA's version of environmental architecture was a response was, as noted, a "war that has been waged against this planet for the last 200 years"; overcoming that war required a professional commitment that could transcend the considerable labor it implied. Thus Yehuda described total commitment as central to making RSIEA faculty work meaningful; its consequences did not end with the individual students who would train there. They would extend to fulfilling her own role in combatting the ecological problems that in her own generation had only worsened. Her work simply followed Mayur's example, she told me, in which responsible environmental work was accomplished only when it was enfolded in one's sense of identity. She described her work as ideally reaching far beyond the classroom, perhaps influencing, if even in a very small way, environmental conditions in India, and perhaps even the broader world. The sometimes-India-specific, sometimes-globally-focused RSIEA curriculum reflected this almost nested sense of the environmental architect's mission: however remotely, it was connected to a global environmental crisis and its appropriate suite of solutions. It also seemed to underline a rather crucial sense of defiant hope, here elaborated as a refusal to be "immobilized by the ugly reality." It was only through such refusal—enacted as ecology in practice—that the environmental architect could maintain the capacity to both envision and operationalize a future of good design.

. . .

The "ugliness" of present conditions was quite real, and the challenge they signaled immense. Mumbai's staggering growth projections, extreme air and water pollution, deep and enduring asymmetries in housing conditions and material wealth, and a multi-faceted but oft-repeated story of urban development in exclusive service of land speculation, coastal degradation, and rampant disregard for human rights or ecological concerns made the very suggestion that architects—or any other collective of urban professionals who sat on the margins of the nexus of urban development power—could have an impact on the city's environmental present and future at best naive and at worst perhaps destructive. My everyday conversations in Mumbai regularly cast doubt on the actual "real world" potential of a collective of architects studying environmental design. How, I was repeatedly asked, could figures *other* than builders, bureaucrats, and politicians influence the development trajectory of Mumbai?

Yet from a social position inside RSIEA's world of good design, I learned to see a persistent and almost dismissive confidence. Obvious and tenacious obstacles—historical, bureaucratic, political, economic, and geographic—could either give way, or were already giving way, to positive change. To my voiced expression of doubt, Dr. Joshi assured me that a real and discernable architectural-environmental change was in fact already happening, but in order to see it, one needed to adopt a historical perspective. With halting optimism, he met my skepticism with a personal sort of evidence, one anchored to a certain view of the arc of urban development in Mumbai:

> What I have seen is that demand (for environmental architecture) is improving, and scope is improving. But you have to learn to see the changes; they may not be what you are expecting. There are many (increments), some of which are very small. Like providing a place for waste processing or using a certain design aspect to improve the quality of ventilation or light. And today the changes are there: people in Mumbai are more aware and conscious of the environment. They are willing to do new things.

Still unconvinced, I asked whether these incremental changes were sufficient to open the kinds of opportunities for practitioners of environmental architecture that would yield both power and autonomy (two things quite clearly missing in the Mumbai of the present). In response, he repeated the entreaty to "see" in a particular way—to notice attributes of change that escape the metrics and usual framings of maldevelopment in Mumbai.

> What is happening is that the number of environmentally certified buildings is very few. But this is because it involves cost and some of the requirements are difficult to meet. However, many builders are including more individual (environmental) features, like recycling water and using it for flushing or for terrace gardens and vertical gardens. Or like keeping more open space. . . . These small aspects are being implemented, but (builders) are not going for certification because it requires lots of things—use of nonconventional energy or energy-efficient equipment, and things like this. At least you can say that 50% of what is required for a green building is being implemented on a regular basis in Mumbai. And environmental architects are very important here.

More importantly, he told me, the scope for future environmental design opportunities was improving because Mumbai's specific urban-cultural sensibility was in part about example-setting, and in part about taking risks. In the context of India, he suggested, Mumbai is a city unlike any other; here, a portion of the public had the capacity to invest in the things they wanted, which increasingly included environmental vitality. He referred, of course, to only an elite subset of Mumbai's wider population, but within that subset Joshi saw qualities specific to the city. This logic placed Mumbai in a powerful cultural and economic position to adopt patterns of environmental improvement as evidence of progress. He explained:

> People in Mumbai have money power. They are willing to spend money and become an example. The culture in this city is entirely different (from the rest of India). They don't bother worrying—if a client in Mumbai decides, "I want this," then they go for it. . . . Outside Mumbai, (most) people are not that ready to spend money. They are more conservative and frugal. Mumbai has got tremendous purchasing power. . . . And also, people talk. . . . They may not go and look and study an issue themselves . . . but whatever a hundred people are talking (about), they will talk about the same thing. And I personally have worked with many projects in which it is clear that mentally people in Mumbai have already accepted this concept of the need for green buildings. And whatever is possible they want to do. . . . So there are good things you can see, and the willingness is there. Clients say, "if you give us the good and foolproof technology and ensure that no problem will be created by this design, we are willing to accept it." The spaces are constrained in Mumbai, but money and willingness are not a concern. The scope for environmental architecture here is good, and it is getting better and better.

For Joshi, then, the same forces associated with the city's urban development dynamics—greed, power, and conspicuous consumption—could be construed as sources of hope; indeed, these could be understood as real-time evidence that environmental architecture was taking root and would eventually flourish.

· · ·

Through convictions derived from a shared, integrated subjectivity and devotion to an integrated subject, RSIEA's faculty worked and taught in ways that both depended upon, and yet quietly defied, Mumbai's urban political economy and development trajectory. Joshi's logic of "money power" even construed the political economic status quo as central to his logic of change; the very force that had driven so much ecological destruction could be imagined as its own undoing. Over time, I came to realize that Joshi was not alone in his full expectation of dramatic shifts, not only in urban development patterns and built forms, but also in the political economy that suspended them in its web.

Through curricular design, shared architectural practice, and modeling an environmental architectural community that was the discernable product of their work together, the core faculty of RSIEA sought to train, inspire, and support the architects who would forge the good design of a better urban future. The environmental architect could not be an expert in every disciplinary dimension relevant to good design, but by learning to see an interconnected system of processes and considerations, she would ideally be prepared to identify, assemble, and critically assess the diagnostic work of socioenvironmental reinvention. This integrative sensibility formed the core of producing design prescriptions worthy of the good design designation. RSIEA's formulation of environmental architecture thus promised to turn the destruction of conventional architecture into the promise of stewardship. Teaching it as a vocation allowed its adherents a sense that they were transforming architecture itself from environmental destroyer to, at the very least, its benign enhancement, and at the very best, that same destruction's direct socioenvironmental remedy.

Ecology in Practice

Environmental Architecture as Good Design

"Most of what we need to learn we can only know from visiting the building site. The rest we can learn from Indian history and a spiritual focus."
—DR. DODDASWMY RAVISHANKAR

"To look at pre-independence buildings is to see sustainable design staring back at you."
—SUHASINI, AUROVILLE

"Take this very seriously, and remember that this whole semester is about values. You're questioning what is right. You're moving way beyond architecture."
—AR. PRITI BANDARI

It was with great anticipation and curiosity that I joined a new cohort of RSIEA students for a formal welcoming ceremony and lecture program in the winter of 2013. The full day agenda began with a film, followed by lectures from several RSIEA faculty, guest speakers, and alumni. In between, we learned some of the everyday logistics to expect from the next two years of student life, but for most of the day, our group of architects-newly-turned-students was invited to contemplate the urgency, purpose, and responsibilities that would accompany a Rachana Sansad degree in environmental architecture.

I arrived at the large auditorium in time to greet some of the faculty members and settle into one of the room's red plush theater seats. I scanned the printed agenda, a bit surprised to see a familiar film title at the top of the program. As new students shuffled into the now packed auditorium, the lights dimmed, a film screen descended, and former Vice President Al Gore quipped, "I'm Al Gore, and I'm the next President of the United States."

The fact that an infamous election defeat erased Gore's presidential aspirations aside, I wondered how a film I regarded as a standard among American environmental studies audiences might come to life in this very different context. At the same time, I paused over the very fact that our first collective RSIEA experience, and perhaps even the framing narrative for the welcoming program itself, was *An Inconvenient Truth*, a 2006 Academy Award-winning American documentary film on climate change. I had viewed, and shown, the film several times, drawing from it regularly in my own environmental studies courses back in New York. But viewing it here, among a group of architects who were now students of the environment, gave it a curious set of new possibilities. Although useful for signaling the global-scale stakes of any kind of environmental training and action, Gore's lecture-driven, PowerPoint slide-laden film struck me as nevertheless awkward and somehow out of place. Perhaps an anthropological preoccupation with context specificity had led me to expect something that was, at least, more overtly architectural and at most, more attuned to distinctly South Asian concerns and imperatives for green design. Regional environmental predictions were dire, after all, spanning issues of future water scarcity, crippling levels of air and toxics pollution, enormous coastal populations vulnerable to sea level rise, and massive and expanding socio-economic inequalities. This is not to say that I hadn't expected program presenters to invoke global climate change, but rather that locating its narrated starting point with a film made famous in part for the role of a former American vice president seemed surprising to me, but quite natural to the audience I sat among.

The film outlined a historical narrative of global environmental awareness that I myself had invoked often when I taught in the U.S. From our auditorium seats, we gazed together at *Earthrise*, the stunning 1968 Apollo Space Mission photo of Planet Earth, followed by the even more ubiquitous *Blue Marble* Earth photo taken from Apollo 17. These images, and the historical moment of consciousness-raising they had come to index, drew us toward an imaginative leap from our physical places in an auditorium in Prabha Devi, and the city of Mumbai, to conceptual scales of larger regions and even larger global landscapes. *An Inconvenient Truth* portrayed just that: a scientifically coherent set of interlocking biophysical systems that were under dire and intensifying stress; these would require dramatic reorientations in politics, economies, and policies to alleviate. We all had a place in the reorientation process: salvaging the global future from the ravages of climate change would involve not only science, but also collective acts of consciousness-raising, environmental stewardship, and decided ecological engagement. From the center of this narrative, it was difficult to differentiate between the global, universalized planetary risk the film emphasized, and the deeply heterogeneous social and geographic texture of the localized threats that climate change posed. To commence our RSIEA experience by focusing on this planetary scale afforded a temporary entitlement to think beyond the messiness of places, including the place in which we sat, transfixed, and watched. When the very future of an aggregated humanity was suspended in the balance, the intricacies of Mumbai's political and social environment—or any of the city's specificities, for that matter—seemed almost a decadent luxury to consider.

The two-year course began, then, with what might be regarded as a conventional, Western-centric, undifferentiated narration of environmental belonging and responsibility: human beings inhabit a common planet, share a common future, and depend on a biophysical context more vast and complex than any scale at which we live individual or everyday social lives. To situate the question of responsibility for having caused, or for perpetuating, climate change across a continuum attuned to historical circumstances and power relations seemed to miss the global point, and its attendant moral imperative.

An Inconvenient Truth surely had another possible effect. By opening the RSIEA program with a film so fixed on the global scale, the faculty conveyed to new students that the curricular agenda would prepare them to assume a legitimate place in global circuits of knowledge, data exchange, and organized responses to environmental change. An RSIEA degree would activate more than locally situated expertise; it would prepare its environmental architects to navigate the global arena of green expertise.

But the lights came up and quickly drew us firmly back into place. A guest visiting professor in that term, Dr. Doddaswamy Ravishankar of the Indian government-owned Housing and Urban Development Corporation, stood before us at the prominent if age-worn podium. Behind him the screen that just moments before had led us to imagine the vast universals of planetary scales now read simply, "The Morality of Sustainability." As if to balance the blue planet image with an equally galvanizing regional narrative, he quickly clicked the keyboard to summon a new slide. It recited a familiar passage:

> I have traveled across the length and breadth of India and I have not seen one person who is a beggar, who is a thief. Such wealth I have seen in this country, such high moral values, people of such caliber, that I do not think we would ever conquer this country unless we break the very backbone of this nation, which is her spiritual and cultural heritage, and, therefore, I propose that we replace her old and ancient education system, her culture, for if the Indians think that all that is foreign and English is good and greater than their own, they will lose their self-esteem, their native self-culture and they will become what we want them, a truly dominated nation.
>
> Lord Macaulay's address to the British Parliament, February 2, 1835[1]

In an instant, India's distinctive historical experience of European colonial expansion, and Mumbai's particular political and economic conditions under the British Raj, flooded back into the picture. Ravishankar's aim was not to ponder the authenticity or ubiquity of the quote, but rather to remind his audience that the anthropogenic origins of climate change were not embedded in a uniform global history of burning fossil fuels; its very emergence depended on extractive and exploitative global political economic patterns that were themselves administered according to specific values and moral sensibilities. The very stuff of Enlightenment notions of progress—the Industrial Revolution, the extractive networks and trade systems through which it expanded, and the modern history of development across the postcolonial world, were all quite unevenly implicated in the global environmental conditions of the present. While at the planetary scale we might all share future consequences quite likely to be indifferent to which communities are more or less implicated in its cause, he suggested, it may also be the case that the political economic circumstances within which climate change was enabled had left us distracted from certain historically practiced or known alternatives. If it was through the colonial and twentieth century postcolonial political economy that the present crisis was forged, then the ecological distortions that came in its wake were inextricably tied to historical processes of erasure, domination, and in Ravishankar's framing, a loss of the very "thing" that made those in the room collectively eligible to claim the identity of *Indian.*

An educational undertaking like RSIEA, one that would draw part of its content from regionally-specific built forms and ideas that long predate India's colonial experience, thus began to situate itself in an explicit accounting of the responsibilities that students might consider distinctly "Indian" or unique to "Indian" cultural identity in an era of climate change. From this vantage point, Ravishankar implied, the environmental architect in India would learn the scientific language associated with the scale of the Blue Marble, but also proceed with an eye toward recovering the situated past; such a recovery was essential if we were to redirect the environmental future. We were in Mumbai, after all, a city whose contemporary built landscape was woven with the remnants of textile manufacturing and the laborer housing that characterized much of the later years of colonial rule there. Beyond the transforming millscape, the city was richly animated by grand, iconic structures that daily retold the history of the city's colonial social and spatial order. The meeting point of planetary environmental stresses and appropriately "Indian" remedies, according to Ravishankar, would thus partly lie in the architectural work of historical reclamation. Though the room was packed with students from a complex array of backgrounds, the cultivation of their collective identity as *Indian* environmental architects had clearly begun.

"These days in India we have money and technology, but what is lacking are the institutional mechanisms to create sustainability," he continued. These were not foreign to India, he assured us; rather they were deeply ingrained in regional history as ancient and indigenous. A sustainable sensibility need not be imported, he lectured: "We need only look to our own past." Repeatedly appealing to the importance of "maintaining our integrity," Ravishankar told students that the long history of "imported, Western ideas of sustainability" exposed it as "hypocritical" and profoundly incomplete. Western building practices denied "a place for the intrinsic spirituality of sustainability. It is Western nations that should be looking to *us* to learn about sustainability; it is only India that can teach them inner growth." Ravishankar underscored his powerful point by invoking this supposedly innate, essentially "Indian" understanding of the intersection of spirituality and sustainability.

With those in the auditorium riveted, Ravishankar narrated a sharp reversal of the colonial calculus of power and dominance: perhaps we were not only here at RSIEA to bring Mumbai, the region, or the country into compliance with a global trajectory that would reverse climate change. Perhaps the promise of Indian environmental architecture was its power to reorient historically dominant moral ecologies as well. Assuring the audience that the foundational ideas of sustainability were present in ancient Hindu texts, he said, "most of what we (environmental architects) need to learn we can only know from visiting the building site. But the rest we must learn from Indian history and a spiritual focus." An implied conflation of "Hindu" and "Indian" continued as a discursive automatic, leaving open the question of whether and how the various origins of those historically "Indian" traits we would study as sustainable would include the region's far broader, more diverse religious and ethnic attributes. But for the moment, the larger point was clear. Training at RSIEA would not involve the uncritical absorption of globally circulating metrics, techniques, or narratives of ecological dysfunction. It would expose students to these, but demand in addition the contextualizing skills to accept, adapt, or reject them as valuable, as "Indian." The stakes were global, but the tactics would be profoundly local. Ravishankar departed to enthusiastic applause, and the group sat chattering long after the house lights came up and summoned us to a tea break.

A bit later, reassembled in the auditorium, another speaker, a representative from Govardhan Ashram, addressed the group. This particular semester, RSIEA would conduct a weekend study tour at Govardhan—a sort of "test run," the program head told me, to determine whether the ashram was an effective location for staging some of the program's experiential curriculum. More details of that study tour and the ashram itself await in a later chapter, but in the context of RSIEA's welcoming ceremony and following Ravishankar's impactful declaration of the role of Indian "spiritual focus" in environmental architectural training, the appearance of an ISKON ashram spokesperson hinted toward a very specific rendering of the form that "spiritual" might take.[2] At a time when India's Hindu Right was gaining political strength and dominance, it was difficult to reconcile the diversity of the new student body and RSIEA faculty with repeated references—both overt and implied—to Hinduism specifically and a conflation of spirituality and environmental thinking more generally. Would contextualizing environmental architecture for "Indians" automatically invoke shades of Hindu nationalism?

The focus on spirituality faded from prominence, however, as a set of lectures focused more directly on retelling aspects of the urban planning and development history of Mumbai. Mishkat Ahmed, an architect and urban planner based in Mumbai, gave a talk that invoked the case of Navi Mumbai, the planned township area northeast of South Mumbai, to explore how development plans might address socioeconomic asymmetries or certain social and environmental questions. She focused heavily on the well-known Indian architect Charles Correa, who played a central role in conceptualizing and advocating for Navi Mumbai. Ahmed used Correa as a more localized counterpoint to the global figure of Vice President Gore, reifying his prominent place in the regional planning imaginary. But she also echoed the previous talks insofar as the emphasis on Correa allowed her to directly relate responsible action vis a vis the environment to "Indian" ideas and practices.

Though Correa's reputation as an architect-activist has been exhaustively debated, the content of those debates was rhetorically less important in this instance than the invocation itself: here was a figure quite familiar to RSIEA students whose efforts in the case of Navi Mumbai could be used to reinforce the idea that environmental architecture and the idea of an Indian moral ecology were logically connected.

In some ways, there was nothing necessarily new in these repeated discursive linkages between Indian "traditional" architectural knowledge, a generalized notion of Indian identity, and socioecological problem solving. There was also little new about drawing on Charles Correa to substantiate such claims. At least since the 1980s, conventional transnational architectural discourse about Indian identity:

> inevitably considers architecture as an agency historically influenced by, and capable of influencing or solving, future social and cultural problems and challenges perceived to be a given in Third World situations. By not so complex translation, hence, such architects are then promoted variously as visionaries, cultural messengers, or as Charles Correa is considered, an "activist" of such a necessary and radical change.[3]

But what was perhaps notable here was the recurrence, in this earliest experience as a collective of RSIEA students, of claims to the transformative, almost agentive power of activating "Indian" identity in the context of contemporary environmental architecture. The vast contents of both would emerge across courses, field experiences, and collective engagement, but our starting point reinforced the notion that responsible environmental design could only derive from a specific, Indian historical rigor. Over time, the curricular case for this would build for individual projects and their pasts, but also for deeper patterns of power relations and social organization. The experiential field visits I detail in a later chapter were a key arena for this.

Invocations of simultaneous global belonging and regional historical specificity thus traced a discursive arc that began with the moral urgency of global climate change but concluded with tellings of the contextual, and even individual, lifeworlds of Mumbai's situated architects. In both starting and closing the opening program, the leap from environmental architect to activist was in fact no leap at all.

. . .

Environmental architecture's moral imperative thus framed, its contents—as the concepts, design techniques, and architectural technologies that constituted "good design" at RSIEA—would be its essential building blocks. In the weeks and months after the opening ceremony, I reported daily to RSIEA to attend classes, travel with students on field study and project excursions, and puzzle over occasional assignments. According to Dr. Joshi, the founding faculty member introduced previously, even as modifications to the curriculum "updated" course content, hybrid teaching strategies and methods were an enduring ideal:

> When we started, (we emphasized) . . . more of the classical things like recycling and reusable materials and how the environment works. Then right from the beginning we (took) students to live projects. And now we want to do this even more. Taking them to places like Auroville, water treatment plants, or to different buildings where innovative materials like compressed earth blocks are used . . . this shows them examples of how it's done and that is central. In fact, in the new curriculum there's more of that now. And we try to give assignments to students that emphasize self study. We give them guidance but we make them study the topics on their own; then they do the presentations to their class of twenty students. If each one takes a topic, they cover lots of ground on their own, and they get presentation practice and experience . . . They also get the confidence to explain the concepts to others. They will need to be able to do this with clients, so they start here.[4]

The "environment" in environmental architecture was thus gradually defined through a combination of problem-oriented field situations, such as how to derive locally-sourced and ecologically sound building materials in a place like Auroville, and the more conventional recitation of lecture and course material that presented a mosaic of knowledge forms from the biophysical sciences, technology and policy studies, and a cultural history of built forms and "sustainability" in India. Experiential field learning on study tours and field projects gave students a chance to try to apply their newly acquired integrated knowledge in practice.

Throughout my fieldwork, the curriculum was structured so that students studied these themes simultaneously; courses might emphasize specific material and proficiencies in science, policy, design techniques, or technology, but most felt deeply integrative, cross-referencing and mutually reinforcing one another as students progressed through them. Two versions of the curriculum are relevant here; as the program was transitioning between old and new syllabi, I experienced both rubrics. From the student's perspective they were fairly indistinguishable, class titles and topical emphases shifted slightly in the transition.

In the original curriculum, the biophysical scientific basis of environmental architecture formed a foundational starting point. Instructors used a systems approach to living organisms, the physical environment, and the flow of energy to convey a basic definition of ecology across four courses—Introduction to Environment and Sustainability, Disturbances and Remedies, Urban Ecology & Environmental Management, and Environmental Services Management Systems. In these, terms like "ecosystem," "ecology," and "sustainable" signaled scaled units of human and nonhuman elements in patterned, usually quantifiable interactions.

Following assumptions generally associated with mid-twentieth century models of, and assumptions about, ecology, these interactions were then assumed to beget defined trophic structures, to reproduce biotic diversity, and to host the constant exchange of materials across units and within different parts of any given unit. In tracing these exchanges, and defining the relevant scaled unit boundaries, ecosystems could be designated—either conceptually or in the practice of engaging a site for architectural design. This approach echoes definitions grounded in Odum's 1953 work, *Fundamentals of Ecology*; its core concepts—like order, mutualism, and cooperation in nature, an assumed trajectory in the nonhuman world toward "balance" or homeostasis, and its focus on communities of living organisms in constant interaction with physical environments—resonate with the implied and overt definition of "ecosystem" that was formally and informally imparted in the four ecology courses and onward through the curriculum.

A definition of ecology derived from this specific version of systems science mapped most directly to the lectures and readings for Introduction to Environment and Sustainability. Here, students read aspects of Odum's *Basic Ecology* alongside several other reference texts, including *Environmental Science: the Way the World Works* by Nebel and Wright and *Modern Concepts of Ecology* by H.D. Kumar. This course introduced and defined "sustainability" as a logical counterpart to the working definition of ecology; drawing from a collection of texts that included works considered to be classics of the mid-twentieth century Western environmental movement (such as Rachel Carson's *Silent Spring*, Jane Jacobs' *The Nature of Economies*, and *The Gaia Hypothesis* by James Lovelock), as well as publications by Rashmi Mayur (the figure discussed previously who had first inspired RSIEA's Head, Roshni Udyavar Yehuda), the course constructed an almost seamless conceptual relationship between functional, vital ecological systems, and "sustainability."[5] In a way that echoed the opening session, it also placed an Indian figure, Mayur, in a prominent place among North Americans often invoked when sketching the mid-twentieth century rise of Western environmentalism.

This conceptual rubric conveyed an interrelationship between "ecology," used as a frame for explaining *how* the environment works, and "sustainability" as a metric of its vitality and value. Human life, particularly in the concentrations and numbers we experience in the historical present, was reinforced as the source of inevitable environmental disturbance, and the challenge to the environmental architect was framed as the mitigation of adverse impact. Maximally functional nature—free of human-induced disturbances—was fully desirable, sustainable, and good in this framing. Here emerges a preliminary guideline for understanding the frequent use of "good design" as the aim of both the RSIEA environmental architecture curriculum and, once trained, its responsible practitioner.

Fundamental biogeochemical cycles like the carbon cycle, water cycle, and nitrogen cycle were covered over multiple courses, and their influence on the built environment, as well as the reverse, conveyed through principles of conservation and efficiency as applied to space, energy, and material resources. These principles were an additional core focus of Introduction to Environment and Sustainability, and they were repeated across the curriculum in terms of ecology as a problem in which "disturbances" must be identified and "remedies" devised.

To elaborate this logic, the course called Disturbances and Remedies was designed as a compliment to Introduction to Environment and Sustainability. Here, the guiding conceptual principles derived from standard environmental impact assessment models; core texts include Canter's *Environmental Impact Assessment*, P.K. Gupta's *Methods in Environmental Analysis*, and Biswas' *Environmental Impact Assessment for Third World Countries*. Following Joshi's narration of the RSIEA program's origins, the focal disturbance for this course was pollution, disaggregated into physical, chemical, and biological expressions. A central concept here was that specific social characteristics can help to maximize ecological vitality, and certain aesthetic, cultural, and social disturbances can distort it. The architect might find those social dimensions difficult to define and clearly problematize, but the physical, chemical, and biological aspects could be measured and managed as air, water, solid waste, and noise pollution. The course introduced students to fairly standard—that is, internationally consistent—technical procedures for remediating pollution. By studying environmental impact assessments and disaster management plans, students were further encouraged to consider their potential role as architects in mitigating pollution, and therefore maximizing "sustainability."

In the following semesters, two courses built on this foundation: first, Urban Ecology & Environmental Management applied the concepts to urban agglomerations. In four units, students study Environmental Problems of Cities, Mobility and Infrastructure, Environmental Planning & Disaster Management, and Urban Hydrology. Reference texts such as *Integrated Land Use and Environmental Models, Cities for a Small Planet*, and *The Gaia Atlas of Cities: New Directions for Sustainable Urban Living* underlined a central message that the inevitable ecological disturbances associated with human settlements in previous courses automatically multiply in scale and intensity in cities. One of the greatest challenges to the environmental architect, then, is environmental architecture in cities. "Urban" was usually used interchangeably with "city," but as the program proceeded, students were encouraged to notice the ways that natural resource flows and movements of labor, capital, and information rendered an urban continuum between city and countryside.

Finally, a course in Environmental Services Management Systems presents techniques for managing water, solid waste, and landscape flora. The emphasis here is a menu of internationally available technologies, but also modes of assessing each technology's appropriateness and feasibility in context. India's specific experiences with technologies considered "appropriate" or "inappropriate" are underlined with reference texts such as Agarwal and Narain's Dying Wisdom: the Rise, Fall, & Potential Wisdom of India's Traditional Water Systems, and a heavy emphasis is placed on decentralized, small-scale techniques like the DEWATS (decentralized wastewater treatment) system used in Auroville and the rainwater harvesting systems most commonly used in southern India.

The reader may notice the dated nature of many of the course reference texts, as well as their grounding in both late twentieth century Western environmentalism and in some of the so-called "Global South" voices that challenged and revised its assumptions of political neutrality and universality. Some texts also mark a place in global debates between environmentalists that have played out for decades, such as the relative appropriateness of rubrics like the Gaia hypothesis.[6] In general, the point of the RSIEA ecology course series was not to expose students to the latest scientific papers on urban ecosystems, or even to elaborate their understanding of ecology principles by introducing the many significant revisions to the science that have punctuated recent decades and continue to change in real time. As a consequence, many basic conventions of contemporary ecology—including heterogeneity, patch dynamics, disturbance ecology, theories of chaos and other historical challenges to the very notion of homeostasis—go unstudied.

It is thus critical to underline that the RSIEA curriculum does not profess to create ecologists or environmental scientists. Its curricular structure does not invite students to undertake rigorous scientific inquiry beyond core concepts of systems, interconnection, and basic energy and nutrient cycling. Instead, the ecology courses are integrated into the rubric that Dr. Joshi called "the big picture;" it organizes particular definitions of environmental stresses, impacts, and problems. Ecology is in this sense closely related to the paired discursive metrics of relative "sustainability" and "good design." Readers should not interpret RSIEA's ecology courses as a pedagogical attempt to teach what disciplinary specialists would identify as the "state of the art" in the dynamic ecology subfield of urban ecosystem ecology. To the contrary, the meaning and content of ecology here was rendered in the curricular experience itself; it signaled a modality of interconnection and unity in which anthropogenic built forms and their socionatural context were expected to produce particular, new socioenvironmental contexts. The goal of "good design" was to minimize "disturbances" and to maximize a generalized environmental vitality. Such an approach to ecology necessarily lifts systems science and systems thinking out of their own temporality and dynamism; in so doing, ecology for RSIEA environmental architects was rendered as a frame with specific diagnostic, relational, and functional attributes. The dynamism of those attributes followed a very different temporality and trajectory than other social renderings and practices of environmental expertise.

The curricular transmission of ideas of a science called ecology to a technical practice called environmental architecture should not be understood as a linear progression from a domain of knowledge, in this case ecology, to its operationalization, in this case environmental architecture. As social practices, or as different forms of ecology in practice, both the science and the architecture involve the production of specific kinds of knowledge, validated and reinforced by their respective audiences. In this sense, it is somewhat misguided to interpret RSIEA environmental architectural practice as somehow failing to incorporate "actual" ecology—a critique that might arise from a perspective that seeks an active operational domain for scientific ecology that is directly connected, in real time, to the changes and innovations always happening in that field. Research in science and technology studies, following the foundational work of scholars like Latour and Jasanoff, has repeatedly shown that fields of scientific knowledge production are also fields of expertise, epistemological domains in which the practitioners of a given form of expertise are technicians; their work involves a constant negotiation between the political and technical spheres.[7] The knowledge that is legitimated in each arena is rendered for, and affirmed by, specific audiences that are deemed qualified. In the case of RSIEA, then, it was within the Institute itself, and in the social experience of training, that the specific form of green expertise called environmental architecture was made and remade, verified and re-verified.

It is also the case, however, that the foundational assumptions and assertions that came to stand for ecological knowledge in RSIEA environmental architecture were derived from a knowledge domain regarded as authoritative and legitimate; the contemporary scientific ecologist might read those foundational assertions and contest whether they are rightly "ecological" at all. They are, undeniably, considered out of date in scientific ecology.

The curricular content at RSIEA thus outlined its own definition of what ecology meant for an environmental architect, identifying the technical details through which her expertise would be assessed, and the audiences and networks to which that expertise would be held accountable. An impulse to distinguish clearly, or draw a fixed connection, between ecosystem ecologists and environmental architects risks losing sight of the distinctly different temporalities, knowledge forms, and legitimizing audiences that shape their practices and compose the networks to which those practices and their agents are ultimately held accountable.[8] The epistemological domain of the "environment" in RSIEA's form of environmental architecture was therefore produced in the social experience of the curriculum, the interactions through which it was conveyed and contested, and ultimately, in its practice as a form of green expertise. In that domain, an environmental architect was assessed by the extent to which she practiced good design, not her expertise as an ecologist, biologist, chemist, or any other natural science discipline.

Of note here is not only the ways that conceptual borrowing between fields can also redefine or temporally freeze the concepts themselves, but also the temporal hybridity of RSIEA's particular form of green expertise.[9] Contemporary green architecture at RSIEA built upon historical notions from ecology, but inside the arena itself, it was precisely those elements that helped to transform conventional architecture to a practice that could take on the challenges of the present.

Having worked through its key ecological content, the remaining curriculum addressed environmental disturbances and mitigation techniques. Ravishankar's opening day assertion that "most of what we need to learn we can only know from visiting the building site" echoed through a strong curricular emphasis on the importance of knowing the experience and physical aspects of a given building site, even if a team of disciplinary specialists might be needed to fully understand them. In specific circumstances, design considerations like building placement and orientation, climatic context, and the availability of recycled or reusable resources were shown to facilitate strategies for minimizing built form impacts; these could be combined with available technological tools to enhance an architect's accomplishment of "good design."

An instructive example of RSIEA curricular treatment of architectural impact and mitigation can be drawn from a session called Green Home Technologies, which was part of the week-long course sequence our group undertook in Auroville. As I will describe in more detail in a later chapter, the annual RSIEA study trip to Auroville is by far the most popular among the several field study programs offered at RSIEA; it played a central role in the experiential reinforcement of many facets of the in-class curriculum.

Auroville hosts a variety of environmental architecture experiment sites, and enjoys an international reputation for a certain kind of experimental architecture. A RSIEA faculty member described the city to me as the "epicenter of sustainable architecture in India"; indeed, the popularity of the so-called sustainability science trainings it offers to visitors attests to national and international renown. This endures, despite Auroville's reputation in other areas as something of a relic of mid-twentieth century countercultural utopian idealism.

The city maintains a close connection to the Sri Aurobindo ashram, and its foundation in the spiritual philosophy of Sri Aurobindo suggests the kind of hybrid attributes to which Dr. Ravishankar had gestured on opening day. In fact, what Ravishankar might call Auroville's "spiritual focus" is a complex product of a historical movement associated with the ashram in nearby Pondicherry. Housing roughly two thousand people, the city was founded in 1968 by the followers of Sri Aurobindo and Mira Richard (known more commonly as the Mother); the latter had called upon devotees to create Auroville in a guise that would allow it to become, as its explanatory literature espouses and its residents repeat, "the city the earth needs." I shall engage this mission more fully in a later chapter, but for the moment let us return to a curricular experience of learning the definition of "good design."

Our instructor for the Green Home Technologies session introduced herself by only her first name, Suhasini—a practice consistent with all of the instructors who led our varied courses and workshops. Trained at the Delhi School of Planning and Architecture, Suhasini is a partner in the Auroville design and planning firm AB Consultants. She is also an Auroville resident, or "Aurovillian." Her welcoming remarks framed Auroville as a generative place "where architects can try things out, experiment, and research."[10]

Suhasini opened her lecture by suggesting that the aspiration to practice environmental design generates a tremendous sense of pressure. A commitment to it seems to imply the need to take on many different goals simultaneously, she explained, and to meet them all in every project. Suhasini cautioned against this, assuring us instead that "it is not necessary to do everything all the time everywhere," adding, "there are certain technologies that are only sustainable in particular circumstances." She offered an example: an architect designs a water recycling mechanism for a building located on a site with a high water table. "This is totally unsustainable, even though it sounds great to say the building recycles its own water," she said. Following this logic, her guidelines for "good" design emphasized that it is sometimes counterintuitive. Rather than trying to maximize the number and types of environmental interventions in a single design, she said, "consider the context and do more with less. If you are doing *these* things carefully, you're on your way to good design."

Suhasini then outlined a clear map of principles for good, or as she continued to call it, "green" design. The first element was "minimize everything." "Everything" encompassed needs, design interventions, and special engineering techniques. "Try not to add to the problem, but rather, be the solution through your intervention." Second, "work in terms of multiplicity of function." Here, guidance centered on maximizing the possible utility of a given space in order to avoid "the unsustainability that comes from the lack of intensive use." Third, "design for all aspects of climate." Suhasini linked this to a concept of "biological harmony" that signals minimal "stress" to occupants inside a built structure.

The next point followed: "design for durability and longevity." Avoid creating excess construction waste, since this is usually dumped in ways that are harmful to the environment. As an instructive example she cited a nearby bird sanctuary that doubled as a clandestine repository for construction debris. "The problems that follow this dumping will be with us for decades to come," she cautioned. Astonishing quantities of PVC and steel lie in heaps across the sanctuary territory, "plus the materials themselves are lost to us. We can't use them once they've been dumped." The desirable alternative is to select materials that use base resources efficiently, and "one way to do this is to choose materials not because they are the easiest to procure or the most familiar to use." To justify this as good design, Suhasini invoked the past, noting that, "pre-independence, materials were procured from a 25–50 kilometer radius; notice that to look at pre-independence buildings is to see sustainable design staring back at you." Noting that different products have different energy inputs and pollution effects during their own production process (the idea of embodied energy), she advised the group to consider the toxicity of new materials and to seize any opportunities to recycle. "We would not be where we are without cement," but its high carbon unit cost makes it one of the most polluting industries. "Beware," she cautioned, "of materials that don't age, show stress, or need maintenance!"

Suhasini continued to offer more precise ideas of "green" design by describing her own style of professional practice. Begin by valuing the ability to cooperate, she said; "the days of the master architect and his minions are lost; none of us wants to be a lab rat anymore. Therefore team playing is essential!" Invoking the past once again, she continued:

> Very often what architects have become is service providers. But historically we are not this. We were people who made changes. We have become the last guys in the pipeline, not influencing clients and users as we should. Remember that architecture is a profession that is more than a service. We have a say, and we have to be responsible for it. We need to be there as projects are being formulated . . . (and use our) position of tremendous influence.[11]

The lecture concluded with a reflection on what the architect can hope to achieve in practice:

> We are not so delusional as to say we will achieve sustainability. We have to design in a way that enhances sustainability. Do this with appropriate built forms, materials that are local, harmony with climate, and the goal of capturing and reusing available resources. Avoid producing hazardous waste. Avoid all waste. Aim to enhance sustainability.[12]

With that, we were left with a buoyed sense of the agentive capacity of environmental architects—on an individual basis and as active participants in cooperative units. Suhasini's elaboration of good design concluded precisely at the point of our potential, one we may have lost sight of in the present, but which, according to her invocation of the past, had strong and inspiring precedent. It seemed unimportant to our group to discuss the structural parameters in which she worked, or the peculiar economic and bureaucratic apparatus that facilitated and oversaw architecture and development in Auroville. The agentive potential Suhasini invoked suggested that all good design can transcend the confines of specific social structures. Obstacles or perceived limits, in this formulation, were no match for good design in practice.

. . .

Back in the Prabha Devi classroom, the broader definition of good design always suspended at least partly in the sociality of its making, RSIEA students move from courses that define the biophysical principles of ecology to those intended to convey a "toolbox" of strategies, technologies, and metrics to supplement their craft. In the new version of the curriculum, the main courses in this cluster are Sustainable Building Design Principles, Sustainable Building Materials, and Thermal Comfort and Passive Design.

FIGURE 4. A team of RSIEA students prepare a topographical map of the Pali field study site. *Photo by the author.*

Sustainable Building Design Principles is organized according to the formal themes listed below; these reinforce previous courses in which ecology was defined and notions of balance, harmony, interconnection, and homeostasis are associated with sustainability. The themes supplement this notion of the relationship between ecology and sustainability by introducing a history of the international, Western, and Global Southern environmental movements of the twentieth century. The course includes exercises in thinking across scales and contexts, as well as the idea of carrying capacity for habitats.[13] The final thematic cluster of the course introduces quantitative approaches to assessing relative building efficiency and the ways these are aggregated to form various international indices of sustainability. The curriculum lists these themes:

1. Understanding the term sustainability: sustainable development an overview of report of Brundtland commission formerly the World Commission on Environment and Development (WCED), Earth Charter and other summits by United Nations. Brief history of sustainability from agrarian communities largely dependent on their environment, western Industrial Revolution tapping vast growth potential, advances in various fields, environmental movement and energy crisis in 20th century, to increasing global awareness—greenhouse effect, etc. in the 21st century, global treaties & action plans.
2. Sustainability principles and concepts- scale and context: over many scales of time and space: environmental, human, cultural, social, technological social & economic organization.
3. Total carrying capacity of Planet Earth; extent of biological and human activity or part of it. consumption-population, technology, resources: destruction of

biophysical resources & Earth's ecosystem, environmental impact, complex
ways in which resources are being used; renewable resources; resource management in economic sectors, manufacturing industry, work organizations etc.
Attempts to express impact mathematically.

4. Measurement: measurements used as the quantitative basis, metrics used
for measurement of sustainability, indicators-benchmarks-audits-indexes,
accounting-assessment and appraisal-measures of reporting sustainability-
environmental sustainability index and environmental performance index.[14]

The experience of classroom sessions and lectures affords another window on the
making of good design expertise at RSIEA. In Dr. Doddaswamy Ravishankar's
winter 2010 course Design Principles, he opened one typical lecture by asking
students to brainstorm how the term "green" applies to building materials.[15] "What
does it mean?" he asked, poised at the blackboard with a piece of chalk in hand.
"Low consumption!" said one student. "Conserve energy," offered another. A third
added, "biodiversity." Lines filled the blackboard: recyclability/biodegradability of
a material; less embodied energy; low emissions/low waste generation; non-toxic.

Pausing the exercise, Ravishankar asked students, "Now, how would you organize these into a green materials protocol? If we have to make this list into something we can use to choose the right materials for green building design, how
would we do it?" The students stared back, some seeming to reflect, while others
were simply puzzled. After a few moments of silence, Ravishankar suggested a
parallel list of questions to guide materials choices:

I would organize it according to a set of questions: what are the exclusions? Meaning,
are there thresholds or laws about the material you're considering? And then, what
are client's preferences; do they desire a more energy efficient building? What about
the benchmarks for all the building inputs and outputs . . . how much water consumption are we talking about, for example, and how does this material relate to our
goals? Now, how about the management system? Is there some way that a materials
manufacturer maintains consistency, like through a monitoring system or certification? What about disclosure—how transparent is the story of how this material is
made? Then you want to ask about the material's compliance with environmental
and social expectations . . .[16]

The instructor then identified a host of international organizations and their
websites; each, he said, offered useful examples of how to organize a materials
protocol. Mapping the world as he composed his list, he encouraged students to
study the APO Tokyo Eco-products Directory. Here they could explore how the
organization mapped the life cycles of various building materials. He pointed to
the German Wuppertal Institute for its database of embodied energy in common
building materials. Coming to the case of India, Ravishankar emphasized the
absence of an Indian standard for, or even clear definition of, biodegradability.
"Here, you run into formal definition problems every time you consider a green

building criterion," he said; "when it comes to India, try to get beyond the criteria and use your common sense."

Building on the materials protocols to which he had referred, Ravishankar introduced the idea of the Sustainability Assessment of Technology, or SAT. These protocols, developed by the International Environment Technology Center of the United Nations Environment Programme, are often considered the most integrative because they incorporate environmental, social, and economic measures of performance and acceptability.[17,18] Claiming a term from green capitalism and marketing, he called these criteria the "triple bottom line."[19] As an example, he raised the idea of "local" materials sourcing, noting that this suggests environmental benefits, but it also might connect to social and economic capacity-building at that same scale. "Beware, though," he cautioned; "never interpret the SAT as a matter of scoring. Use it to remember that linking materials or technologies to social well being is always important."

The world map of examples continued. Ravishankar introduced the international Environmental Products Declaration system administered from Stockholm, and encouraged students to visit the website of the American Institute of Architects, as it had just held an important exhibition on embodied energy in building materials.[20] The list of transnational protocols continued to grow. The International Environmental Technology Center of the UNEP offered a useful consolidated fact sheet on materials, while the national materials rating systems in Germany and Austria had "some of the best rating systems." As the class period came to an end, the blackboard was scribbled with lists of websites, international protocols, and examples from elsewhere, and the promise that by the end of the semester students would develop their own grasp of the menu of materials available to them, and a wide range of approaches to assessing the relative sustainability of each. Most importantly, they would be able to develop their own guidelines, appropriate to the Indian case, particularly drawing from, as Ravishankar stated, their "common sense."

In addition to familiarizing us with a vast array of considerations when it came to building materials choices, the protocols exercise had the effect of conceptually reconnecting our aspirations to practice environmental architecture with a wider transnational community of institutions and their associated metrics. Charting the many tendrils of a global movement called environmental architecture—here, by mapping the contours of its protocols—only reaffirmed the importance of good design expertise on the global scale. With the Indian context as our anchor, we were nevertheless guided across the global landscape of guidelines that shaped good design expertise beyond our sphere of practice. Here was another dimension of the hybrid knowledge form being forged in the social experience of training: our expertise was affirmed at multiple scales, and the need for it was as global as it was "Indian."

Just after this lecture, my hand sore from scribbling protocols lists in my notebook, I moved along with the students to the next course, Thermal Comfort and

Passive Design. Shirish Deshpande, our lecturer, greeted us and launched directly into a deceivingly simple question. "Which is greener," he asked, "GRIHA or LEED?" These two metrics for assessing the relative sustainability of a given built structure automatically posed an Indian protocol against one that circulates internationally, and as such, is sometimes regarded as a global standard.[21]

Deshpande paused, and then began his response. "Any code that is pushing toward a new baseline—that is continually pushing the envelope, so to speak, is good." Again, the case of India became the exclusion: he cautioned that LEED requirements are based on very specific models that often depend on data and product availability not immediately applicable to India. "These are standards developed in the U.S., so naturally they are not always appropriate for India," he said.

But the question had opened another point, to which Deshpande devoted much of the remainder of the class session. "You know," he began, "It's better not to look at the credit systems only; look at the intent, and start with your design. What do you want to achieve?" Environmental architects must follow a conscious design process, he explained, not just proceed according to scoring from a list of points or credits. "But Sir," a student replied, "even if the client is not asking for a green building, we can design it in this way." Yes, was the reply; *this* is good design. "If you forget about the credits and just think about good design you will automatically have a good building. And you will teach the client that it is smarter than racking up points." Good design, then, demanded an agentive stance: it implied both the capacity and the responsibility to "teach" the client, and it demanded that the architect work with, but move beyond, the procedures that followed from protocols and codes. Practicing good design was not following a recipe; it demanded, in fact, the opposite stance.

The central theme of Deshpande's session that day was the importance of using standard metrics as "tools for checking and reference," but never as guidelines. To elaborate, he gave an example: differentiating between COC (costs of construction) and OC (operating costs), Deshpande supposed that a "typical Mumbai building" carries normal construction costs across a range of INRs 1,700–2,000 per square meter. A student quickly offered that "building green" would increase the cost to at least INRs 2,400 per square meter. "Can we reduce this first cost?" Deshpande asked. "How can we capture and convey to the client that there is a payback in the long term? If we start from the notion that the first cost of good design will always be higher than conventional design, then we will never do green work."

Despite what seemed to be an obvious impasse born of economic realities, the instructor pressed on. "Be creative," he urged. "What if you, as the architect, just give it a (cost) cap? What if I say design a green building and keep the cost down to INRs 1,200 per square meter? Could you do it?" The room was silent. "It's up to you, the architect. You can take the lead in making good design decisions." No obstacle, he implied, even an economic one, should be stronger than good design.

In both sessions, Ravishankar and Deshpande ascribed almost infinite potential agency to individual Indian architects armed with good design; the moral imperative then rested with each student to simply perfect and employ it. To fail to do so signaled, in large measure, the weakness of the architect alone. The suggestion that one can design one's own materials protocol whilst proactively consulting a full range of international metric tools was at the core of Ravishankar's blackboard laden with lists, while Deshpande left students with the clear directive "It's up to you." If a process seemed too expensive, the architect could make it affordable. Metrics and protocols were heuristics rather than guides. "You can take the lead," Deshpande assured them, and the power seemed to rest with the hybrid expertise of good design. Suhasini's lecture back in Auroville was not only echoed in the classroom, then; it was squarely reinforced. Recall:

> Remember that architecture is a profession that is more than a service. We have a say, and we have to be responsible for it. We need to be there as projects are being formulated . . . (and use our) position of tremendous influence.[22]

A final example to supplement those drawn from Auroville and the Prabha Devi lecture hall can be drawn from the tours of green building sites that were included on our various study tours. On an RSIEA visit to green architecture sites in Chennai, the first was the India corporate headquarters of Grundfos Pump Manufacturers. The building was India's first-ever LEED-certified gold building; at that time, "gold" was the highest LEED rating yet awarded in India (though this would change almost immediately afterward).

At our first stop after several hours of driving, the group streamed out of the bus and filed through the building's front entryway. Some paused, taking note of the prominent plaque directly to the right of the entrance. This was the marker that certified the building's gold LEED certification. At its center was a cluster of leaves, and written in large print around this image were the words, "U.S. Green Building Council" and "LEED," along with the certification designation—in this case, the gold medal.

A lingering student soon turned to me and, in an ironic tone, made explicit the obvious question that I imagined we pondered collectively. "Why does the US Green Building Council determine what is sustainable in Chennai?" Why should appropriate parameters for defining sustainability *here* come from *there?* "It's not ecological," the student said; "the climate and the materials—and really everything—these are different from construction in the U.S. A gold building there is not sustainable here."

Grundfos Headquarters itself, which for our purposes was a study site, was simultaneously a material link to transnational circuits of sustainability definition and assessment, *and* a staging area for formulating and articulating claims about, and the stakes of, precisely what good design is and how it is enacted. As we walked its corridors and observed its features, the building was itself a kind of provocation to define good design in place and time, an invitation to assess the sustainability achievement signaled by gold certification.

As the formal tour unfolded, aspects of the meaning of, and work performed by, gold certification revealed a particularly corporate inflection that up to that point we hadn't engaged in classroom lectures. Just beyond the doorway, a greeter addressed our group in a large, light- and plant-filled foyer. One wall listed the Grundfos official code of conduct; alongside it was a comprehensive list of the building's green technical features. Together, they professed deep commitment to a version of sustainability that embraced a specific kind of eco-capitalist morality, one that echoed many of the standard principles of so-called green capitalism.

The tour itself was a highly stylized and technology-savvy presentation. In addition to its environmentally sound design features, our guide told us, the building also made a positive "social contribution." Laborers here did their jobs in the most light-filled and fulfilling of spatial settings, which in turn produced, he claimed, "much happier, more productive workers." Here, "social" good was assessed directly as increased production, and quite clearly, higher profit for the company.

Our visit eventually led to the office of the regional CEO for Grundfos, who elaborated more fully the specific transnational corporate culture within which the "message" of the building was nested. The Indian Regional CEO of this Danish company spoke from an office desk decorated with miniature Indian and Danish flags; behind him an enormous Danish flag draped across an entire wall.

The CEO opened his remarks by assuring us that Grundfos is "very profit oriented." But, he continued, "the important question is *how* we generate that profit. We don't want to be number one; we want people to know us by our commitment to the environment."[23] The company's mode of conveying that commitment involved the global vocabulary of sustainability, namely USGBC LEED certification. The LEED gold building stood for the presence of certain types and numbers of technical features, but also for its place in an international corporate geography of a specific kind of capitalist commitment.

The students were especially interested in the design features that had "earned" this building its LEED gold status. Each time the tour guide identified something, small groups seemed to join around the feature and discuss it. The tension between our guide's narrative emphasis on "scoring points" and the students' general desire to understand why certain design approaches were chosen over others eventually produced an almost palpable unease, exacerbated when the guide explained that most of the strategies employed to maximize LEED certification points were those that were easiest to undertake. At every instance, the design team avoided approaches or materials that would alter the cost, challenge conventional materials mixes, or dramatically modify standard construction and design practices. Grundfos had simply gathered all of the proverbial low hanging fruit, and the result was the visibility enabled by LEED gold status. After all, we too were there at the headquarters.

As we completed the tour, our guide moved swiftly between pride in a kind of moral achievement and pride in "getting the gold" without exerting much effort.

In blatant contrast to the modality of good design emphasized in the RSIEA class-room, the stewards of this building seemed to mark environmental architecture attributes only when they turned on questions of profit: energy efficiency saved the company money, for instance, and worker productivity boosted profits. This sort of "green" building was a rather transparent, strategic capitalist strategy rather than an example of good design. Several students later labeled the tour, with dis-gust, as "greenwashing."

At the lunch break, Professor Rajeev Taschete and a group of students talked through the part of the tour we had just completed. A spirited conversation heard students listing item after objectionable, and often absurd, item. "This is just a checklist," one student said; "it's not good design at all. It fails in all the ways that matter! If this is the example, LEED gold means nothing for India."[24]

The tour of India's first LEED gold building was not, then, a study of how to follow in this champion's footsteps. On the contrary, it unfolded as a systematic critical exercise in which nearly every built form aspect that earned LEED points was debunked as somehow contrary to good design. In practice, then, neither the "follow the protocol" approach, nor the outside metric, lived up to the standards of good design.

. . .

Emboldened with the hybrid knowledge form derived from classroom lectures, and the confidence in their agentive potential reinforced across the curriculum, students who reached the last part of the RSIEA program enrolled in a capstone-style course called Environmental Architecture Studio (also called Design Studio). Offered in three consecutive parts, the three courses assembled under the rubric combined classroom time with a design project assignment undertaken in student teams. Part lecture, but mostly field-based, this course gave students their sole opportunity to practice good design under the guidance of RSIEA faculty.

Design Studio courses posed a design challenge that came from an actual cli-ent, and the proposals that student teams created were presented to that client at the end of the three course sequence. In 2012, the brief involved designing a set of villas for an art resort development at Pali, about 40 km from Popoli, and a nearly two hour drive from Mumbai. The students were directed to design twelve resi-dential villas over 6.5 acres of land that we were told was "undeveloped." An agent for the developer visited the classroom as our course got underway to convey the resort owner's vision.

Introducing herself as Shilpa, the middle-aged, fashionably dressed agent described the developer as "young, adventurous, and (wanting) to change the typical attitude." He envisioned a resort that would provide an "escape to nature," she told us; he believed that "when someone comes from the city and gets a natu-ral experience, it changes the state of mind." To further frame the context, she asked students to imagine a setting in which the surrounding villages grow rice, mango trees populate the hills, and "you feel nature." The terrain was rich with boulders, red soil, basalt, and "lots of trees." On a separate, but adjacent, lot, other land owners would eventually construct forty private homes, making the site a "nature escape" that was quickly transforming, becoming more and more a visibly connected node in an urban-rural continuum.

The resort was to be called Serenity Villas; in their Design Studio assignment, the students would divide into teams, each developing a plan for twelve cottages of three to five hundred square feet. A swimming pool, bar, and restaurant were also planned for the resort complex; an amphitheater would be built at its center.

Little was said, or asked, about the people who were already living in the villages surrounding the assignment site, or the land use and land tenure conditions that preceded the making of the Serenity Villas plot. Shilpa emphasized the cityfolk who would journey from Mumbai to patronize the resort instead, describing them as "people in fancy cars who want to experience nature," and "people who are thirsty for nature but they don't know how to enjoy it." As her project description concluded, we learned that the following week's class meeting would take place on the site in Pali. We should plan to walk the land and conduct our first site assessments. This was it: our newly gleaned, hybrid knowledge form would be put to experiential test. The room was giddy with excited chatter as Shilpa bade us farewell.

The faculty coordinator for the Design Studio that semester, Professor Priti Bhandari, had herself trained at RSIEA years earlier. She had gone on to practice environmental architecture in Mumbai with some amount of success. Once Shilpa had departed, Bhandari turned to us with a kind of urgent sincerity. This project brief, she said, would be the culmination of everything we had done at RSIEA. Short of our independent thesis projects, the Design Studio project was "the most important work" we would undertake. If the developer was convinced by our designs, he was likely to actually *build* one of them, so of course we needed to "take this very seriously." Then, as if to underline that good design implied far more than protocols, templates, or metrics, she said, "and remember that this whole semester is about values. You're questioning what is right. You're moving way beyond architecture."[25]

Our next course meeting found us assembled at the site in Pali. The instructor divided us into teams, each charged with a set of data collection tasks that the class could aggregate into a full social and ecological contextual picture of the site. This would be our first experiential attempt to derive the "integrated knowledge" that we would need to undertake good design. A sense of the breadth of biophysical and social data we expected to consider by this advanced point in our training can be gleaned from the assignments delivered to students in that first field visit. Working independently in their respective teams, each was charged with preparing a presentation for the other students on one element of the comprehensive site assessment below. Expectations for the depth and sophistication with which we

FIGURE 5. RSIEA Design Studio students explore the Pali project site.
Photo by the author.

would complete each list were minimal; what mattered was assuming its attendant, expansive view—viewing the site of our design brief in a fully integrated way.

· · ·

The RSIEA curriculum reflected its primary objective: to impart to architects the capacity to understand the environment as an integrated subject. From the opening day lectures through the design studio experience, we learned that environmental architecture in the form of good design could never be reduced to the mastery of prefabricated tools or metrics; it depended instead on developing proper values and an agentive stance not only to conceptualize a practice that could transcend existing structural limits like costs or codes, but to actually do the same in practice. Good design depended on a hybrid knowledge form that was both globally sanctioned and anchored to being "Indian." It assumed few limits to architects' agentive potential.

Through RSIEA training, an active and shared notion of architecture's environmental object was constructed, defined, and translated into a modality of responsible practice. We found that environment, in its fullest sense, in places far flung from Mumbai: from the Pali study site to the tours I will discuss more fully in chapters to follow, environmental concepts gleaned in the classroom were reinforced quite afield from the dense human presence and built development of the city itself.

The vast terrain of political economic and power differences that the good design practitioner would traverse, and the extent to which the architect's work—even as an environmental architect—might reinforce or exacerbate its inevitable forms

PALI Site Visit Assignment

	ASSIGNED TASK AT THE SITE	EXPECTED OUTCOME OF ASSIGNED TASK
Group A: Site, History, Climate, & Air Environment	Site map; historic events; disaster proneness; climate analysis: net and field measurement; air quality and noise parameters measured or net searched; micro-climatic and/or diurnal variations, if any. (All these as researched or collected from site to be incorporated into the sheet as drawing or a summarized/tabulated format.)	**SHEET 1:** Formatted on A1 size: Maps-regional, contextual key plan and site plan as drawing; history; climatic charts with the analysis mentioned therein; polluting activities and their anticipated impact on air environment.
Group B: Land, Topography, & Built Forms	Land holdings; population occupying the land/using its resources/ maintaining for resource; seasonal variation in land-usage, e.g., weekly bazaar/festival. (All these as researched or collected from site to be incorporated into the sheet as drawing or a summarized/tabulated format.)	**SHEET 2A:** Formatted on A1 size: Land-use map categorized for tenure, contour plan, site sections, sketches & photos of indigenous & contemporary construction and recordings of current developmental features.
Group C: Hydrology	Water sources & resource-to- agriculture/people/etc.; sewage and storm water drainage/indigenous methods/etc.; Problems, say water-logging/ seasonal scarcity of water/etc.	**SHEET 2B:** Formatted on A1 size: of the watershed, catchment area, & river basins of the region; hydrology of site (village); water sources; usages; etc.
Group D: Soil, Geology	Collection of soil samples from identified key locations, natural features dependent on the geology; detritus (soil organisms); importance of the natural features to the natives; (All these as researched or collected from site to be incorporated into the sheet as drawing or a summarized/tabulated format.)	**SHEET 3:** Formatted on A1 size: Soil analysis maps, sketches or photos of soil erosion control measures implemented at site.
Group E: Flora & Fauna; People & policies	Document plant associations observed or researched; decline in any species observed in last decade; list of number & types of trees, shrubs, plants, undercover, characteristics & habitat (location). List of fauna sighted/heard/ net search characteristics & habitat. Observed flora-fauna associations NGOs operating in the region and their focus activities; tribal & inhabitants-culture, livelihood, etc.; tribal and land policies-development rules of the region; changing trends in livelihood.	**SHEET 4:** Formatted on A1 size sheets: Vegetation plan, regional and site plan **(village plan digitized from Google)** with trees; data tabulated–biodiversity in photos & pictures, food resource, analysis of the produce, season, and land usage. **SHEET 5:** Formatted on A1 size sheets: Photos & pictures; organizational diagrams in terms of how the NGOs work; notes with references; site observations on activities and usage of resources, space, etc.

Students are expected to:
Make a checklist of parameters to measure and observe at site.
Make checklist of things to carry for site visit for their specific topic.
Make sample questionnaire for any field data required from local inhabitants and/or clients.
Refer to books in the library to develop the format for data documentation for the specific topic.

of social exclusion and violence, were not the direct focus of this curriculum or its attendant praxis. Still, as was suggested in students' strong reaction against the "greenwashing" version of green design we encountered at Grundfos, these issues were inescapable. Much of the time, however, learning good design meant equating proper practice with the almost automatic byproduct of a simultaneously sustainable city and more harmonious society. The precise contours of the bridge between sustainable city and sustainable society seemed both presumed and, at least in overt curricular terms, omitted, but the responsibility to forge that bridge rested unquestionably with the architect properly equipped to practice good design.

4

Rectifying Failure

Imagining the New City and the Power to Create It

Did you know that Mumbai has four rivers?
—OPEN MUMBAI EXHIBITION

Greater Mumbai cannot survive as a concrete jungle.
—BREATHING SPACE EXHIBITION

As the students trained to practice good design, parts of the wider city of Mumbai were caught up in a wave of events, symposia, exhibitions, and spectacles that amplified anticipatory optimism for the city's new development plan. This chapter offers the reader one trajectory—the author's—through a selection of public events that highlighted the possibilities for sustainability suggested by the imminent plan. In doing so, I aim to pause and rescale our focus from the social experience of Institute training to the wider city and its publics. These, after all, form an important dimension of the broader social worlds that all of the students lived among. This chapter addresses the social production of ideas of good design as they were nested within a wider urban frame for the potential place, and composition, of Mumbai's environment. As in the context of the Institute, that frame was produced in real time, its aesthetic and ecosystem service dimensions promoted, contested, and reworked across many specific publics and locations in Mumbai.

I note from the outset that each event I discuss in this chapter indexes specific attempts to influence the extremely complex and layered world of Mumbai's urban politics.[1] Yet, unlike the RSIEA context, this chapter does not attempt an exhaustive analysis of the events or the publics they created or excluded; while these are important issues worthy of their own book-length analytical treatments, they are beyond the scope of the present work. I note them here because their frequency and presence in this period of my fieldwork reinforced in more popular settings the sense of purpose and urgency signaled by the RSIEA concept of good design. Like me, students would daily come and go from the Institute in Prabha Devi only to pass by, deliberately attend, and often take part in, the events I describe below. They help us understand, then, the wider social climate within which environmental architects were situated—one characterized by the active making of self-designated Mumbai publics who deemed specific social and natural transformations necessary to salvage the city from otherwise inevitable socio-ecological chaos. By highlighting a set of public events, this chapter then proceeds to a question that is central to forging a link between any form of green expertise and engaged social practice: namely, who, precisely, controls urban development in Mumbai?

I proceed, then, to recount a subset of the many spectacles through which the ideal future city's contours were defined, debated, and mapped in the lead-up period to the new development plan. Each was organized by a different group—and therefore enabled different kinds of claims to legitimacy—and each took as its central concern not the question of *whether* one or another vision of an open, greener Mumbai was desirable, but rather which version could be understood and embraced as the most appropriate, representative, and, ultimately, sustainable.

A few weeks after my arrival in Mumbai, I was seated among an audience of Mumbai-based planners, architects, and urban professionals in what, like the RSIEA opening program, might be viewed as both a global and a postcolonial setting. In the quite Victorian assembly hall of the former Victoria and Albert Museum, now the Chhatrapati Shivaji Maharaj Vastu Sangrahalaya, the group had convened for an event called Reimagining Mumbai. Organized by the city's Urban Design Research Institute (UDRI) and faculty and students from Harvard University's Graduate School of Design, the seminar drew its participants from Mumbai's environmentally interested urban professional public, as well as an internationally mobile, elite group of graduate students and their mentors. The challenge, as the name suggests, was to undertake a daylong, collective visioning exercise: what would a greener blueprint for the city's built form—one that would reorganize its land-use mosaic while meeting the needs expected to accompany a dizzying future population growth scenario—look like? The program comprised speeches from Indian and international "experts," but its singular message was that regardless of one's home context or degree of familiarity with Mumbai (indeed, some speakers were visiting the city for the very first time), all assembled were somehow entitled to register their voice toward the goal of reimagining it.

I was invited to this gathering by the RSIEA Head herself; she was among the urban professionals who had participated in discussions, projects, and preparations that informed the event. As such, I'd expected her work or her presence to be in the relative foreground, but instead the program included some presenters who seemed to have little or sometimes no prior experience working in Mumbai. I sat in the audience, then, alongside Udyavar Yehuda, someone who in the RSIEA context would be leading the group.

The program was eclectic. Some speakers offered comments that seemed to reinforce what by 2012 had become a rather standard, globally circulating narrative of Mumbai, informed by its iconic status as a planetary epicenter of informal and slum housing.[2] Without question, the challenge of re-housing, or differently housing, the estimated 8.7 million people living in Mumbai's slums is both critically important and notoriously tenacious, but international discourses of slums in Mumbai at that time sometimes had the effect of allowing slum settlements to stand, in singular dimensionality, for the city more broadly.[3] At times this betrayed the complexity of social, political, and material life in slums themselves; it also supplanted more careful attention to the layered social, political, and biophysical challenges the city faces as a whole.[4]

Among the audience gathered for Reimagining Mumbai, invoking discourses of Mumbai as a city of slums certainly underscored the importance of any attempt to reimagine the city's urban landscape; it also placed the focus on the architect's very brief: namely, shelter. As panelists discussed the city's development plan, they rehearsed a dire picture of the present, but connected it to new possibilities. Urban growth projections foretold a massive future buildingscape that did not yet exist but would rise quickly. Between 2012 and the next common demographic benchmark year, 2025, the UN estimated that Mumbai's population would grow to 27 million, a sharp contrast to the city's official 2011 population of 18.4 million.[5,6] This scenario added urgency to the task of forging a plausible urban material and social fabric; the Mumbai of the present was already famously failing to provide basic services like safe housing, adequate infrastructure, and sufficient reliable utilities.[7]

It is further important to notice the distinctive spatial politics of twenty-first century environmental projects. Not only is an aspiring urban environmental architect guided by analytical questions of identity formation (the "we" of similarly trained and value-oriented architects) and tellings of history (defining *Indian* green architecture), but she is also guided by a somewhat constant scalar code switching between globally circulating metrics of appropriate environmental architectural forms and those considered to be uniquely anchored to scales more accurately defined as "of the place." The converse is also true. As environmental sustainability increasingly figures in urban aspirations for relevance on a global map of cities that "matter," the environmental architect may be regarded as having the potential to hold the power to foreground specific cities on a global stage.

If the iconic global image of Mumbai is so dominated by the idea, aesthetic, and human experience signaled by global imaginaries of "the slum," then the chance to refashion Mumbai's housing into more ecologically desirable forms not only suggests a chance to rectify the majority population's unlivable reality, it also potentially assigns environmental architects a central role in any effort to "reform" or refashion Mumbai's global image.

In addition to housing, the panelists emphasized a second urban failure, one that placed Mumbai's biophysical stresses at its center: flooding vulnerability. In 2005, the eighth heaviest rainfall ever recorded for Mumbai had caused catastrophic floods.[8] At least 5,000 people died, countless others lost their homes, and thousands were stranded in life-threatening circumstances on Mumbai's roadways. References to the floods—signaled in the simple mention of the date, July 26— punctuated public events surrounding the new development plan and underscored the city's urgent need to be "open" and "green." At Reimagining Mumbai, the director of the city's Urban Development Research Institute called July 26 a "game changer, because," he declared simply, "on this date the city failed." Environmental catastrophe—the record rainfall—combined with the calamity of decades of specific patterns of landscape transformation, had exposed the city's distinctive and inexcusable vulnerability to complete socioenvironmental breakdown. The city could not stand by and wait for such events to be repeated, panelists argued, reinforcing the imperative to reimagine Mumbai.

While some of the environmentalist alarm over demographic growth projections have rightly been understood through the analytic of "bourgeois environmentalism," those same projections also bear on a biophysically grounded, ecological understanding of the damage that decades of water and infrastructure development patterns have wrought in Mumbai.[9] As Rohan D'Souza and others have shown through their critical approaches to hydrology management and patterns of development in India, it is only through attention to historical alterations of Mumbai's morphology, watershed, and cityscape—in the Mumbai case, often in the form of landfilling to facilitate urban development—that we are able to see how crucial hydrological patterns may have shifted over time, concentrating and intensifying flooding incidents in specific areas, and ultimately perhaps preventing the city from coping in extreme storm events.[10]

Reimagining Mumbai was one of a full range of productions, exhibitions, and spectacles convened in this period, each enabling specific, sometimes highly localized and institution-specific conversations about the possibilities of urban salvage through landscape redesign, city "greening," and aspirations that nearly always reinscribed the urgent mission of the architect and urban planner. As social spectacles, those arenas were simultaneously spaces for delineating specific publics and their audiences and for establishing ideas of appropriate metropolitan engagement, civic duty, and entitlement to undertake the active envirosocial stewardship that would be needed to forge, and follow, whatever the future path toward a reimagined Mumbai might trace. In the event I describe briefly below, the idea of creating and multiplying "open spaces," be they parks, protected zones, or newly vegetated landscapes, was an essential component of realizing a more desirable future Mumbai; as such, open spaces—however they were defined in each event—were regarded as one of the most important attributes of a responsible new development plan.

The events themselves had peculiar parameters. Both the participant pool and related, intended audiences drew from largely elite, relatively young, and otherwise privileged subsets of Mumbai's population. Those who would "reimagine Mumbai" did not formally represent its majority, and at times did not even include its citizens, but they formed a collective category of citizens, government officials, urban professionals, and international "experts" enacting their concern for the environmental future of Mumbai. Embedded in that group were many who possessed or aspired to possess new forms of green expertise, and many who understood and supported the general contours of the social, cultural, and environmental realities such expertise sought to enact. The sphere derived its legitimacy in part from the spectacles themselves, which often stood for evidence of broader "community consultation." Each produced an affect that affirmed that open spaces were widely valued and supported. Despite being civic engagements, however, they were rarely, if ever, predicated on a wholly inclusive notion of or consultation with the full range of Mumbaikars who stood to gain or lose from the future form of the city or the open spaces its green experts so adamantly promoted.

A useful way of thinking about open space promotion through public events is as a mode of cultivating the city, which K. Sivaramakrishnan and I discussed in *Cities, Towns, and the Places of Nature: Ecologies of Urbanism in Asia*. In that work, we noted that:

> parks and recreation, often twinned concepts, visually evoke the idea that a city must enfold nature within it, and provide amenities to the modern, civil city dweller to afford time, quite literally, to breathe in the park. Cross-culturally, the urban park represents plant life, birds, green vistas, clean air, and the uncluttered, protected space in which mind and body can be united, children can play, and refuge can be taken from the daily grind of city life.[11]

Although Reimagining Mumbai was temporally bound, its idealized connection between specific characteristics of urban space, in this case open space, and a socioecologically vital city was neither exclusively postcolonial nor exclusively modern. Sivaramakrishnan and I pointed to Ali's extensive work on Indian gardens from the Buddhist to Mughal periods, among others, as a treatment of this issue in premodern green spaces;[12,13] we further noted that urban traces of colonial rule are evident in the very form and location of existing "green" amenities like gardens, parks, zoos, botanical collections, and greenways in Mumbai. Across colonial cities, planners often sought to combine precolonial urban forms with Victorian ones, thereby forging modern urban natures as a kind of postcolonial inheritance.[14]

Thus Mumbai on the verge of a new development plan marked a period within which specific postcolonial, self-appointed publics cultivated notions of desirable urban open space and the polity that should attend it. To participate in their public events was also an affirmation of entitlement—indeed, duty—to do so. We may in this sense extend the idea of "cultivating the city" to cultivating civic green expertise.

OPEN MUMBAI

A few weeks after the UDRI/Harvard event called Reimagining Mumbai, a provocative exhibition opened at the city's National Gallery of Modern Art (NGMA).

FIGURE 6. An exhibit-goer ponders a map of Mumbai's open spaces at the *Open Mumbai* exhibition. *Photo by the author.*

Open Mumbai was a multi-media spectacle of future urban design ideas, all rendered by the Mumbai-based design firm P.K. Das and Associates.[15] It met with such enthusiastic success that its initial NGMA run, from March 15–April 7, 2012, was extended; it was then re-extended for a month at the city's Nehru Center.

The project on which the exhibition is based, also called "Open Mumbai," was an extensive, ambitious urban plan and proposal for regulatory reform. Originally intended to be exhibited at the Sir J.J. School of Architecture, the first Indian institution associated with the Royal Institute of British Architects, the move to the NGMA signaled in part the popular purchase of long-term visioning exercises in this period. Various extensions of the exhibit only underline this point. The new venue allowed more public access, and offered the audience a multifaceted, multi-media performance of translating a highly technical urban development proposal to a public-civic spectacle of the possible future city.

The exhibition was organized as a multi-tiered walk through the exhibition hall, and on my first visit, I found that walk to be fully consuming. Traversing exhibit sections simulated a journey across the various urban landscapes that made up the firm's "open space" rubric for Mumbai. Enormous, encompassing mural-maps outlined the city's many and varied neighborhoods, while also typologizing its spaces—as they are in the present, and as, through a highly guided exercise in imagination, Das and Associates argued they could and should be in the future.

Consider the exhibit's introductory text, which is also found in the introduction of its accompanying book:

> As Mumbai expands, its open spaces are shrinking. The democratic 'space' that ensures accountability and enables dissent is also shrinking, very subtly but surely. The city's shrinking physical open spaces are of course the most visible manifestations as they directly and adversely affect our very quality of life. Open spaces must clearly be

the foundation of city planning. An 'open Mumbai' ensures our physical and democratic well being. Unfortunately, over the years, open spaces have become 'leftovers'
or residual spaces, after construction has been exploited. Through this plan, we hope
to generate dialogue between people, government, professionals . . . and within
movements working for social, cultural, and environmental change. It is a plan that
redefines land use and development, placing people and community life at the center
of planning—not real estate and construction potential.[16]

As the exhibition laid out the social and ecological components of "open space," it
also made a clear argument for the impossibility of separating the two. Social and
ecological transformation would have to happen simultaneously if Mumbai was
to reach the firm's goal that it "go beyond gardens and recreational grounds—to
include the vast, diverse natural assets of the city," while at the same time fostering "non-barricaded, non-exclusive, non-elitist spaces that provide access to all
our citizens for leisure, relaxation, art, and cultural life." This plan, so the exhibition declared, "will be the beginning of a dialogue to create a truly representative
People's Plan for the city."[17]

This would be no minor undertaking, so the exhibition space engulfed its visitors in appropriately massive maps. "Before" and "after" photos from previous Das
and Associates projects assured the viewer of the plausibility of "Open Mumbai,"
even those aspects that seemed improbable and fantastic.

The exhibition guided its visitor across the vast ecological terrain of the cityscape:
seafronts; beaches; rivers; creeks and mangroves; wetlands; lakes, ponds, and
tanks; *nullahs* (drainage canals); parks and gardens; plot and layout recreational
grounds; historic forts and precincts; hills and forests; city forests; open, people-
friendly railway stations; roads and pedestrian avenues; and area networking all
comprised major exhibit stations. Some of these, such as the parks section, were
clear elements of the city's biophysical space, while others, like "people-friendly
railroad stations," mapped a complex concept of ideal civic society in a specific
design. The scale, color, and consuming details of each element were enhanced by a
soundscape that mimicked what one might hear in each place; one section featured
birdsong, while another repeated the rumbling sound of rolling coastal waves. This
was the soundtrack to the biophysical aspects of Mumbai.

The exhibit experience offered its viewer reassurance, and even evidence, that
the comprehensive vision of Open Mumbai could be realized through the right
combination of urban design, civic advocacy, and governance. While the exhibit
itself did not focus on the latter's complex details, the entire undertaking's
core objective, made clear throughout, was to accomplish "necessary amend-
ments in the DP (development plan) and accompanying DCR (Development
Control Regulations)."[18] The legal apparatus that structured urban development was acknowledged, then, but the chasm between the aspirations depicted
in an exhibit space and their enactment in the actual city was left otherwise
unproblematized.

FIGURE 7. "City Forests" were highlighted among the many different types of open spaces in Mumbai at the *Open Mumbai* exhibition. *Photo by the author.*

The case for "Open Mumbai's" exercise in urban aspiration greeted exhibition visitors, and so framed the intentions for visitors' overall experience:

> Open spaces reflect the quality of life in a city. In India's financial capital, Mumbai, rapid development and expansion of the city has resulted in the erosion of its open spaces at a rate that is truly alarming. The 'lungs' of the city, like recreation grounds, parks, and gardens, along with invaluable natural assets like mangroves, wetlands, forests, rivers, creeks, and the natural coastline, are fast shrinking. It is our opinion that the situation simply has to change.[19]

Thus prepared, the visitor first met with the seafront ("With 149 km of coastline and seven interconnected islands, Mumbai is a city on the sea"[20]), then moved through detailed accounts of each element in turn. In most cases, proposals for the possible cityscape were underscored by examples of previous projects.[21] Some came from the firm's completed work in Mumbai, while others invoked global icons like New York's High Line or Berlin's Spree riverfront. Immersed in Mumbai, but occasionally gesturing to a world of presumably "world-class" urban open spaces, visitors were invited to travel across notions of time, space, and the possible urban landscape of Mumbai all at once.

In a manner that resonated with learning ecology as an "integrated subject" back at RSIEA, portions of Open Mumbai were dedicated to a basic lesson in principles of biophysical ecology. Sections related to seafronts and beaches, for example, included primers in elementary coastal dynamics and drainage. Lessons

in beachfront conservation and coastal nourishment accompanied gorgeous renderings of future seafronts that were at present little more than solid waste dumps.
Even aspects of the historical ecosystem, virtually erased in the present by urban
development, were reinvigorated:

> Did you know that Mumbai has four rivers? Mithi, Oshiwara, Dahisar, and Poisar
> (are) together 40.7 km long. Almost invisible to the city's population, these rivers are
> waiting to be 'discovered,' protected, and their shores revitalized as public spaces.[22]

The exhibition's visual narrative linked biophysical lessons with social objectives;
nullahs, once redesigned, for instance, could host pedestrian and cycling pathways
(precisely 81.4 km of them, according to the exhibit), and appealing renderings of
park-like promenades and pathways reinforced this sweeping potential. Even the
city's most notorious environmental problems—like the morphological distortion
and pollution along the Mithi River, or the progressive degradation of the Sewri
wetlands—could be remedied. Dismal photos of these places in the present were
juxtaposed with green, inviting, and imminently redesigned renderings. In these
moments, the promise of Open Mumbai seemed nothing short of a gleaming eco-
social reawakening, a literal city transformation.

Rather than rehearsing an extensive review and critique of Open Mumbai, my
purpose here is to gesture toward the ways the exhibition articulated an explicit
and constant coupling of biophysical and social transformation, made possible
only through the intentional creation and provision of more open, green spaces in
the city. The exhibit showcased the potential power of urban designers to provide
their constituency with the contents of a more ecologically and socially vital urban
future, and it reminded its visitors that the imminence of a new development plan
for Mumbai gave the whole exercise particular urgency. The form of green expertise invoked by PK Das and Associates promised to remake material space while
offering coupled environmental and social vitality.

Yet even as it mapped a highly detailed and explicated vision of Mumbai's eco-
social future, the exhibition claimed its firm commitment to developing a "People's
Plan." This seemed a blatant contradiction, but it signaled reflexive awareness that
Open Mumbai could be easily dismissed as a top-down undertaking at best, a
savvy business strategy at worst. To fully convince its audience, the initiative had
to gesture toward civic agency and inclusion, even if the actual and active input of
a broader public could be held in valid question.

Across contemporary urban South Asia, scholars have repeatedly shown how
the making of urban green spaces—particularly parks, but also more general
modes of environmental restoration—tends to deflect attention from their sometimes violent, and often socially exclusionary, outcome.[23] In such cases, enacting a
"clean, green city" has repeatedly obfuscated new and not-so-new forms of mar-
ginalization and displacement. Open Mumbai made explicit attempts to mark and
reject such forms of exclusion by framing an ecologically sound city precisely as a

critique of the social marginality and exclusivity that had intensified in the wake of urban growth. At the same time, the social sphere included in the making and viewing of Open Mumbai was extremely selective, and hardly representative of Mumbai's vast public.

Urban environmental spectacles such as Reimagining Mumbai and Open Mumbai simultaneously espoused civic participation, promoted social inclusion through the provision of more green spaces, and promoted a particular kind of civic green expertise. The rubrics they invoked framed open and green spaces precisely as the appropriate *antidote* to social exclusion, even as their channels and audiences were exclusive subsets of the broader public.

BREATHING SPACE

I turn to a final spectacle, consistent in mission with Reimagining Mumbai and Open Mumbai, but standing in contrast as well. Breathing Space, organized by the non-governmental organization CitiSpace (Citizens' Forum for Protection of Public Spaces),[24] Tribe@Turf (a subgroup of Royal Western India Turf Club members), and Hathautee (which called itself a "cultural platform" and was more an Internet platform than a physical group with constant membership), was a two-day semi-public festival to celebrate Mumbai's existing open space and to advocate for new open space provisions. An extensive exhibition of panels, art installations, and craft vendors, as well as two day-long programs of lectures and panels made this more of a festival than a singular exhibition.

Like Open Mumbai, Breathing Space was held at an elite city venue that is also a spatial artifact of Mumbai's colonial past: Worli's Mahalaxmi Racecourse. One motive for using this twenty-three-acre site was to raise awareness that its lease faced imminent expiration, but its use profile also made for a somewhat ironic stage for open space advocacy. Although technically considered an open space, a combination of security guards, strict social norms of use, and the somewhat erratic schedule that dictates when it opens to the public at all (legislated in large part by the horse racing schedule, as during races the track is closed) make this a difficult space to access. When not in use for racing, the racecourse attracts individuals and families who use it for walks, jogging, and other sports activities. During several stays in nearby Worli, I enjoyed regular access and use; I also observed both subtle and unsubtle modes of social exclusion when those who could not claim or feign a place of "belonging" there were swiftly denied access.

As with Reimagining Mumbai, I learned of Breathing Space through the Head of RSIEA. She invited me to join her, and together with a colleague from the Sir JJ School of Architecture, we headed to the racecourse after RSIEA classes ended one evening. We shared a taxi for the short distance between the Institute and the racecourse, but our cab could only inch its way through rush-hour packed traffic.

FIGURE 8. Entering the *Breathing Space* exhibition at the Mahalaxmi Racecourse. *Photo by the author.*

In the idling taxi, I asked my colleagues what distinguished Breathing Space from the many other similar events in the city that I'd already attended.

The answer was quick, and invoked Open Mumbai for contrast. This exhibition, our colleague explained, was based on "action research," in part the product of long-term citywide advocacy by the NGO CitiSpace.[25] Breathing Space would feature installations about each of Mumbai's twenty-four wards, all represented in panels that would each highlight one—perhaps among many—open space issue that the residents of that ward (or presumably those residents who interacted with CitiSpace advocates) identified as important to them. My colleagues preferred this more intentionally consultative model, explaining that Open Mumbai could easily be dismissed as an advertisement for the design firm PK Das. They agreed that Breathing Space organizers made a more genuine attempt to report their research findings rather than a presentation of a finished and stylized development proposal. If achieving a "people's plan" for the future of Mumbai's urban landscape was truly an objective, Breathing Space exemplified a consultation process more likely to assemble one.

In fact, the successive ward-panel groups that characterized the CitiSpace exhibition of its research created a distinctive visitor experience. In contrast to Open Mumbai's guided path through biophysical spatial categories carefully curated with images, texts, and soundscapes, Breathing Space visitors were invited to wander the municipal and social category of "wards." They were free to linger and move about at will, invited as well to areas devoted to other kinds of information tables, food and craft vendors, lecture panels, and—quite strikingly—to simply

FIGURE 9. *Breathing Space* exhibit-goers explore signboards about open space in each of Mumbai's twenty-four Municipal wards. *Photo by the author.*

notice being at the racecourse, immersed in outdoor "open space" itself. The light, sound, air, and dirt of the semi-forested event area, its wide, dusty racetrack in the distance, merged the event about open space with *an experience* of open space. One traversed the exhibition and pieced together its messages at one's own pace, and according to one's chosen trajectory.

On arriving, my colleagues and I were greeted by various acquaintances; several people active in the city's urban planning and design communities had also thought this a good evening for Breathing Space. We made our way directly to the ward-by-ward exhibition, where we found rows of visual panels—narrative mosaics of text, maps, photos, and satellite images. Stationed nearby each ward's cluster of interpretive panels was a CitiSpace worker or volunteer, ready to discuss the content of each board, and eager to offer more detail, context, and guidance.

Data reported throughout the exhibition were drawn from a comprehensive survey, completed by CitiSpace in 2011, of six hundred reserved open spaces across twenty-four Mumbai wards.[26] Standing before my first panel, I asked the woman stationed there to tell me more about the project.

She explained that each panel was devoted to one issue that citizens living in that ward identified—by no means the only issue, but something that "gave a glimpse of what's happening" throughout Mumbai. Each ward was presented in light of a profound but well known contrast: official allotments of open space as legislated by the previous official development plan bore little to no resemblance to the map of *actual* open spaces on the ground. The old plan was not a reliable blueprint, then, for determining the active mode of land use, nor was it a tool for locating actual open green spaces. "There is a terrible mismatch," the exhibit volunteer told me, "between the ground reality and the development plan; we rely on citizens to tell us, day in and day out, what's happening." She explained that a large part of CitiSpace's mission involves advocacy; once an illegal use of a reserved open space is identified, the group assists in filing Right to Information (RTI) petitions and, when appropriate, in moving court cases forward.

Breathing Space emphasized ward-by-ward derived accounts of the actual land use profile across Mumbai in order to make the case that at present, the urban development plan was of little use for charting the city's future. The failure of the plan, according to the exhibition's promotional material, led to a wider failure—a "fatal ratio" of persons to open spaces throughout the city:

> Mumbai is a city of remembered open spaces. "There used to be a playground here" and "Where have all the trees gone?" are laments one hears more and more. All too often, the culprits are the very guardians of these assets and we, to whom these persons are responsible, lack the knowledge or the will to act. Saving our parks, gardens and open spaces is not "elitist" or an aesthetic quest. It is protection of health and well-being—physical, mental, social and emotional—of all, especially the most vulnerable: the under-privileged, the young and senior citizens. On this score, Mumbai is in crisis. India's National Building Code lays down that there should be at least 4 acres of open spaces, accessible to all, per 1,000 population. Mumbai has less than one-hundredth of that: 0.03 acres. Transforming this fatal ratio is a challenge each of us must accept. Greater Mumbai cannot survive as a concrete jungle.[27]

In contrast to Open Mumbai, Breathing Space relied less on cultivating a collective imaginary of the future possible urban landscape, and more on amplifying accounts of mis-allotments of open space. This placed development plan monitoring and enforcement at the center of its mode of cultivating civic green expertise.

Like Open Mumbai, Breathing Space sought to mark and reject the social exclusions that often accompany efforts to expand urban open spaces. In these cases, the ecologically sound city was invoked and mobilized precisely as a critique of the social marginality and exclusivity that had intensified in the wake of urban growth, even as the very specific public which invoked it was itself quite exclusive. It was a lack of open and green spaces, and a more general lack of environmental sensitivity in Mumbai's urban design, that had exacerbated the socioeconomic disparities of the present, according to this narrative.

Classic urban anthropological studies such as Caldiera's *City of Walls* have sensitized scholars to the paired struggles over demands for social inclusion that accompany urban democratization and the material and spatial modes of enacting exclusion that often arise in response. Yet Mumbai at the cusp of a new development plan witnessed an urban environmental arena that simultaneously espoused civic participation, promoted social inclusion through the provision of more green spaces, and forged a kind of civic green expertise. Here, open and green spaces were rhetorically crafted precisely as an *antidote* to social exclusion, even as its channels and audience were an exclusive subset of the broader public. Urban nature was made meaningful, in this sense, for its attributed power to create equity and inclusivity across public life; at the same time, it was assumed to facilitate greater environmental sustainability and urban "well-being."

Although Breathing Space did more to point toward the bureaucratic, corporate, and institutional apparatus that operationalizes the city development plan than did Open Mumbai, both were predicated on the idea that the public assembled for these events, however exclusive in the context of the entire civic body, had real power to achieve the goal of more green and open spaces. Left in the background was the necessary and detailed discussion of how the cityscape of the present could be physically transformed according to the expressed will of that civic body, however exclusive or selective.

. . .

Throughout 2012, dozens of exhibitions, from highly commercialized showcase-style conventions intended to display "green" building products or promote and debate specific practical techniques and metrics, to those oriented toward awareness-raising, public education, and advocacy, took place across Mumbai. The outlines of events above suggest just some of the contrasts, issues, and assumptions that characterized urban greening advocacy as Mumbaikars anticipated the new development plan and a future of environmental and social stresses.

Despite the veneer of unprecedented newness, however, neither tensions between architectural design and elitism, nor the relative civic nobility of imagining Mumbai's future or tracking its present, are necessarily new or novel. In his *House, but no Garden: Apartment Living in Bombay's Suburbs, 1898–1964,* the historian Nikhil Rao offers important reminders that attach public fervor over remaking the city to a much deeper history of creating the peculiar urban landscape that is Mumbai.

To take just one example, Rao describes the inaugural speech given at then-Bombay's 1937 Ideal Home Exhibition. Delivered by Prime Minister B.G. Kher, it focused not on the homes of the future—the point of the exhibition—but on urging architects to work to solve the vexing problem of housing the city's desperate poor. Rao discusses the Director of the Indian Institute for Architecture's response:

> (The Director) acknowledged that the finer details of room design might seem grotesque '... to those less fortunate ... who live in squalor and dirt, not to say filth.' Yet such experiments as the exhibition were necessary, he argued. Through their 'carefully worked-out rooms,' they allowed visitors to know 'what to look for and what to demand when the subject of a *home* is in question.' The planners of the exhibition thus sought, through a staging of ideal domesticity, to *actually work out the parameters of the home* for all Bombay's citizens, not just for those few who could afford bath fittings by Garlick and Company or steel furniture by Allwyn and Company or Godrej Boyce and Company.[28]

If ecology is sometimes shorthanded as "the study of home,"[29] we might read a similar tension in the events given the most descriptive attention in the previous section. One presented its assembled public with ideas of what they should desire, while the other emphasized existing, already vocalized desires. Both placed the appropriate "place" of open and green city spaces at the center of ideas of the ideal urban home.

Although couched in narratives of the unprecedented, in fact the tension between advocacy-based and more elitist exhibitions captured in the contrasts between Reimagining Mumbai, Open Mumbai, and Breathing Space returns the focus to the social production of good design and green expertise as knowledge forms and ecological practices—whether expressed projects of social justice or as broader exercises in imagining and representing the possible. Here, in an arena filled with claims about the necessary presence of a specific kind of space—open space—and its specific characteristics—socioecological integrity—lies the continuation of that aspiration to finally and "actually work out the parameters of home." The urban home of the present deemed unacceptable, and the future nothing short of potentially catastrophic, those parameters and their environmental dimensions were undeniable aspects of contemporary civic responsibility, or, as I suggested previously, cultivating civic green expertise. Designers and architects who had mastered good design and citizens who mastered civic green expertise might then claim a place in the work of transforming the social and biophysical city by reshaping its urban material stage. Whether and how those experts were actually empowered to do that remained largely unaddressed.

AGENTS OF "THE MESS": IMAGINING CHANGE VS. THE POWER TO ENACT IT

If the timely civic task of imagining a new development plan created an excited air of possibility, the contemporary state of Mumbai's urban development governance regime offered its sobering remedy. Perhaps redefining urban socionature and bringing it more fully into the city was possible in the future, but in the present, those who wished to access such a reality could only "escape" to it.

Long before the sun rose one Friday in June of 2012, I climbed bleary-eyed into the back seat of an SUV, the guest of two colleagues, both RSIEA faculty members. We promptly headed off for a weekend trip outside Mumbai. Our destination was their small plot of land in Ali Bagh, a coastal area 150 km from the city and worlds away from the population dense, open space-starved metropolis of Greater Mumbai. My hosts were eager to walk the land they'd purchased many years before, and that they were now slowly, painstakingly clearing and cultivating. Their core ambition was to eventually design a small, ecologically sensitive home so they could stay at what they called their "patch of paradise" for a few days at a stretch.

For diversion during the long journey, my hosts brought a fresh, thick stack of morning newspapers. We surveyed the front pages, and the same lead story repeated over and over: the day before, a disastrous fire had swept through the Maharashtra state administrative headquarters, the Mumbai Mantralaya. I read aloud for my companions this excerpt from the *Hindu Times*:

> A devastating fire swept through the upper echelons of Maharashtra's government offices in Mumbai, killing three people, destroying important files and documents, completely gutting the offices of the chief minister and his deputy, and possibly delaying the under-fire government's plans to overhaul the state's creaky urban infrastructure and housing projects. The fire, which started around 2:45 pm on Thursday June 21 afternoon, roared uncontrollably for more than a few hours, terrifying office goers in the vicinity and throwing traffic on the arterial Madam Cama Road, which leads on to the scenic Marine Drive, in total chaos.[30]

On hearing this, one of my hosts interrupted. "There go the records on the Adarsh Housing Society scam!" The other replied, only half joking, "It's a good thing we're escaping to nature!"

If Mumbai's floods and an unenforced development plan had come to stand for environmental failure, Adarsh exemplified the opacity and deal-making that distorted so many of the bureaucratic processes that legislated urban development. Specific informal processes, often technically illegal but rarely adjudicated, ensured that the profits and other benefits of construction in the city accrued to those with specific political and economic ties. Despite outrage over flagrant violations of building codes and laws, or the involvement of government officials at very high levels, in some ways the Adarsh scandal's audacity seemed to simply exemplify the banal reality of regulatory practice in Mumbai's urban development sector.[31] Public outrage over the scandal eventually compelled the Municipal Commissioner to bring hundreds of construction projects to a halt, citing building code violations and demanding compliance. In response, Mumbai's builders threatened to strike. The back-and-forth posturing was in some ways less important than the complex of officials, builders, and financiers it involved. These were the powerful, if dynamic and often elusive, entities that held the most significant control over Mumbai's present form and future urban landscape.

Mumbai's complex of urban development regulatory bodies, and their roots in previous colonial, state, and municipal bureaucracies is perhaps best traced from 1896. At that time, the Bombay City Improvement Trust was inaugurated, largely in response to the devastation of the plague. Before this, the city's primary urban development goal was to attract residents, so both land offers and building activity proceeded on terms that favored property holders. Regulatory leniency became precedent, and eventually the norm, making later attempts to impose new policies very difficult. The City Municipal Act of 1888, for example, had laid out an enforcement regime for a basic set of building regulations (height, ventilation, certain street widths, setback lines); it also restricted the height of new buildings to 1.5 times the width of the street on the site. Reflective in part of norms and precedent, however, the Act contained no provision for the height of older buildings. Owners of these simply went on making unregulated additions.[32] Those subject to the new regulations, meanwhile, often simply ignored or worked around them.

In its quest to use urban planning as a tool for promoting ideas of public health, sanitation, and connective mobility in the city, the Bombay City Improvement Trust joined in a somewhat tense relationship between two prior authoritative bodies: the colonial entity, the Government of Bombay, and the Bombay Municipal Corporation, or BMC, composed largely of Indian landlords, industrialists, and wealthy merchants. In its first decade, the Trust undertook a sanitation mission; it built worker housing and connecting roads to re-pattern settlement and movement across the city.

In the following decade, however, thinking about appropriate schematics of urban development whilst planning for Mumbai's expansion led a new Trust chairman to propose a revised approach to meeting the city's housing and public health objectives. In 1909, he suggested an "indirect attack" on the housing problem by developing residential building stock in what would become new suburban areas in Mumbai.[33] In this scenario, building outward to the suburbs represented the chance to design new landscapes that promoted public health. These suburban landscapes contained elements like carefully placed roads, parks, trees, and a host of environmental amenities associated with promoting human health. Building Mumbai's suburbs, then, had a great deal to do with promoting a "healthier" urban environment.

Rao describes debates within the Trust that actively questioned the relationship between various environmental elements—landscaped boulevards, parks, and other open spaces—and public health. There was clear agreement, however, that certain spatial characteristics connected (or enabled the circulation of) urban social life and nature. This was thought to actively create human social and physical well-being, according to Rao:

> Orr (chairman of the Trust) was careful to present the boulevard as part of a 'strong indirect attack on the prevailing unsanitary conditions which are largely due to overcrowding.' The suburbs were presented not as an escape from the crowded city for the privileged classes, but rather as part of a comprehensive citywide strategy to attack the problem of overcrowding in the older parts of the city by attracting some residents northward. Kent (chief engineer of the Trust), meanwhile, attacked . . . suggestions . . . that the amenities of the boulevard—benches, trees, strips of green to separate different functions, and so on—should be relegated to the parks that were also envisioned in the schemes. Kent argued that such suggestions were 'based on the false premise that boulevards and parks fulfill the same purpose, and it may be as well to compare the two and examine the points of similarity and difference.' While the park and the boulevard might have some features in common, parks were essentially

static places. They provided the 'lungs' of the city, and they were also pretty to look at and an 'attractive amenity to those living within easy reach.' Boulevards, on the other hand, were of greater importance than parks since they constituted the channels through which 'the life of the place must pass.'[34]

The exchange reminds us that elements of urban nature may or may not "count" as nature, depending on their placement in urban space. Furthermore, environmental elements and amenities have a long history as tools—contested, at that—in urban planning; not only were they understood to have the potential to positively impact public health, but by regarding them as the city's "lungs," a consciousness of their place in broader biophysical processes that interconnect with human social vitality was already present.

Political and economic changes after World War I had important effects on Mumbai's urban development mechanisms, as well as on land prices and housing policies. Through the 1919 Government of India Act, the colonial state significantly reduced its oversight of Mumbai's urban development and governance; in its place, ministries elected by a provincial legislative council were granted power to oversee local self governance and public works.

Around this period, the place of architects in urban planning and development also assumed active prominence. Although the Trust had employed architects to oversee the continuity of design in its undertakings for some time, in most cases developers found ways to tailor projects to their own preferences and financial advantage. As a result, despite their presence, architects had virtually no influence; the only exception to this was in the case of large public buildings.

In the 1930s, the BMC endorsed the Trust's policy of requiring architect-certified drawings for new construction. At the same time, suburban development schemes assumed a new and powerful momentum—the dimension of Mumbai's early urban growth and development history eloquently recounted by Rao. A new generation of Indian architects, trained at the Sir JJ School of Architecture in Bombay, came on the scene as participants in an urban planning process and turned their attention to creating Mumbai's suburbs.[35]

However, observing the rise of the Indian architect, Rao alerts us again to their limited agentive role. He notes as emblematic a debate that played out in the pages of the *Times of India*, in which a former Trust engineer described how the architect's potential contribution to urban development was thwarted. He wrote:[36]

Now the average Lessee of the Trust has very great confidence in his own ability in house planning and very little confidence in architects, good or bad. He only employs them because he must, and when possible he confines them to drawing up plans under his direction and obtaining sanction for them from the Municipality and the Improvement Trust. As regards the construction of the building he, if possible, has an arrangement with them by which they give advice when asked and pose as his architects when trouble arises owing to bad work. He pays them as little as he can and is quite ready to employ an incompetent architect if he thinks him capable to getting the plans passed and consequently capable architects have to cut their fees in order to make a living and reduce their supervision.[37]

This excerpt suggests that professional roles bore little resemblance to actual practice or agentive power in the translation of building projects from design to implementation. In fact, this early twentieth century description of the function and constrained power of architects in Mumbai's development still resonates, a point to which I return below.

By the post-independence period, a new constitution outlined a land, housing, and urban development administrative apparatus largely within the purview of state-level governments. By allotting resources through a series of Five Year Plans, however, the central government also exercised enormous influence. In contemporary Mumbai, a state-level urban planning department oversees the Maharashtra Town Planning Department, the Urban Development Authority, and provisions for water and sanitation services. A Housing and Special Assistance Department exercises authority over housing policy issues, land ceilings, rent levels, and, in the Mumbai case, a suite of often volatile and always foregrounded slum redevelopment policies.

At the same time, several key statutory urban development and planning bodies operate at the municipal level. The Greater Mumbai Municipal Corporation (BMC/GMMC) is perhaps most important among them; it exercised considerable authority up until the mid-1980s. It is the BMC that is technically responsible for the city's Master Plan, including enforcement of related development controls and regulations. The BMC also oversees public transport and electricity provision in the city.

In 1992, the Indian Constitution was amended to provide more authority at the level of urban local self-governance. The Mumbai (earlier Bombay) Metropolitan Development Authority (BMRDA/MMRDA), created in 1975, gained new regional planning powers in the Mumbai Metropolitan Region, a spatial area that combines the city and suburban areas which include Thane, Navi Mumbai, Ulhasnagar, Mira Road, Vasai, Virar, Bhayandar, Bhiwandi, Karjat, and Alibaug. The MMRDA's contemporary jurisdiction is therefore quite vast; beyond Mumbai city, it encompasses nearly four thousand additional square kilometers and a 2001 population of nearly six million.[38]

Contemporary Greater Mumbai consists of two separate legal entities: one, the Municipal Corporation of Greater Mumbai (MGCM), oversees most of the services one might associate with municipal responsibility: water supply, sanitation, property taxes, building regulations, public health, roads, and education. It operates through an Administrative Wing, which answers to the state-level authority of an Indian Administrative Services (IAS) officer who is appointed by the state government of Maharashtra. More generalized authority is exercised through the body's MGCM Deliberative Wing, which is controlled by 227 locally elected and 5 nominated councilors.

Greater Mumbai's second legal entity oversees revenue collection and a host of other governmental functions. Divided between the Mumbai City Collectorate and the Mumbai Suburban District Collectorate, its jurisdiction extends to two city regions. The City Collectorate oversees the old island city of Bombay, while the Suburban District Collectorate does the same for the suburbs.[39]

One cannot overemphasize the complexity of an urban development apparatus with uneven reach across city, state, and federal agencies. Each has its own jurisdictional boundaries and dynamic politics. Likewise, one cannot responsibly imagine Mumbai as bounded by conventional mappings; its built form, coastal zone, and underlying waste, water, and energy systems are all managed simultaneously, but often without coordination across the multiple scales of government that may have relevant jurisdiction. To think of Mumbai, urban planning, or the social and political ecology of the region in terms of a singular "city," then, is nearly impossible in this sense; it demands a considerable disaggregation across space, scales of power, and historical and contemporary distributions of resource and policy control.

Nevertheless, the two agencies most often invoked, and indeed most directly involved in Mumbai's development planning, are the BMC and the MMRDA. Yet this is not to say they are widely regarded as effective. In his talk at the Reimagining Mumbai session described earlier, Pankaj Joshi called the BMC, "defunct without support or capacities," and the MMRDA "at best a financial institution." Also periodically powerful in specific issues related to planning and development is the Town Planning Department of Maharashtra. Crucially, the Municipal Commissioner is appointed at the *state* level, the implications of which are that Mumbai's new development plan—open space and all—would have to be approved at the Maharashtra state government level.

Thus in sharp contrast, perhaps, to the optimism and sense of the possible generated in exhibitions like Open Mumbai or Breathing Space, my interlocutors who worked inside the urban development bureaucratic apparatus often narrated a general sense of "chaotic development" in Mumbai. They emphasized the opaque political climate, the deregulation of the Indian economy, and a similar deregulation of the built environment over the past two decades in the city, arguing that all have converged to transform the economic, sociocultural and material fabric of Mumbai.

An instructive example may be drawn from my conversations with an urban reform activist, planning scholar, and professor, whom I shall refer to here under the pseudonym Laxmi Deshmukh. In one of our discussions about the actual location of power to remake Mumbai's urban landscape, she cautioned that both the failure of the city's previous development plan and anticipatory spectacles that sought a new urban vision for the coming plan had to be situated in a realistic understanding of the bureaucratic apparatus that operationalizes city policies. In some moments, she noted, the existing development plan was enacted and enforced, while in others it was not. She encouraged me to focus on the process rather than the contours of the plan:

Now what happened in the last 20 years is that the economy has gone into a big swing. Mills have closed down, offices are coming up, the IT sector has grown only in this period. Services are growing, banking and everything. The complete character of society has changed from industrial areas to services. Now this 20 year plan, which was prepared 30 years back, actually was prepared in 1989—so for 25 years the plan is valid, until 2014. In this 25 year period huge changes happened yet the plan was very rigid. If I have a factory and it's closed down I can't do anything with it—I don't need a factory—my operations have closed and now I don't want a factory; I want an office building. . . . That is not possible with the rigid development plan. So how do I make way for this? . . . The old structures which are not doing good business can't be pulled down, and the old businesses can't close.[40]

Much of the energy of development plan-focused public events seemed to derive from assumptions that citizen awareness could produce action, and that action could influence Mumbai's development trajectory toward a "greener" city. Yet the complex map of regulatory bodies, and the simultaneous involvement of central, state, and local governance regimes and dynamics certainly configured the agentive power of any mobilized public, to say nothing of urban professionals themselves. To bring aspirations for urban ecology into Mumbai's actual experience of development, then, would likely have to directly involve a host of bodies, jurisdictions, and political figures that were almost entirely absent from the pubic spectacles that encouraged reimagination.

In a series of interviews, Dashmukh described her experience and knowledge of that administrative landscape. Sitting together in her office in the spring of 2012, we began by talking about how the Adarsha scandal, the same issue that had animated the newspapers on my earlier-mentioned "escape to nature" drive with my colleagues, reinvigorated public outrage over disconnections between development legislation and development practice. She began:

For any construction project, people have to give consent. . . . When you apply for building permission, ideally the building plans should go to the fire, water, electric, and sewage departments. . . . The engineers in these four departments have to give clearance for the proposal . . . and a fifth is the overall building permission for the structure. These are all under the BMC; this is a wholly elected body with its own administrative setup to give clearances, except for electric. . . . But what happens, as we understand, is that the building permission is given without (securing the clearances). Proposals are sent, but no comments are received, or comments are received and ignored. So BMC is approving projects without having undergone the approvals. So (consider) the scam of Adarsha: . . . no one bothered (to check the permissions), so a thirty-story building is constructed without the fire authority knowing. Or, he might have given his comments that (would have prevented) this (but they were ignored). So basically most of the cases where building permissions are (granted) the actual required coordination and approvals doesn't happen.[41]

But perhaps as important, if not more so, than the dynamics of oversight and regulation, Dashmukh emphasized again the nested scales of authority that control urban development in Mumbai. Our conversation shifted toward public excitement about greening the city through a new development plan, to which she replied that the most powerful authority in decisions about urban planning was often situated at the state level. In important ways, then, regardless of the political will expressed at Mumbai's municipal scale, it was the political climate in the state of Maharashtra that exercised the most concentrated power over the city's future form. She explained:

Look at Mumbai. You hardly find any green space. Everything is covered with slums or buildings or non-cessed old buildings. Whatever is happening as far as construction is all brownfield development: reconstruction of old buildings or construction of new buildings. This development has to be governed by a certain law, (which) is (the) Maharashtra Regional Town Planning Act. This is overarching, applicable for (the) whole state of Maharashtra, including Mumbai. In it, development plans have to be prepared by a local body, but all the (related) permissions . . . are under the MRTP act. . . . So, if I am a city, if I prepare a development plan, that's not the end of it. It has to be approved and stamped by the state government.[42]

Invoking the example of Mumbai's mill lands—those left unconverted to glittering malls now famously mostly defunct, often dilapidated, and certainly located in parts of the city ripe for redevelopment, Dashmukh then turned to the problem of authority and autonomy at all levels. One outcome of bifurcated state and local power was that most development undertakings simply underlined and exacerbated tensions between local autonomy and other scales of bureaucratic authority. She continued,

(So consider that the) mills closed around 1987, following the strike that started in 1984. They were taken over by the central government, and the land and everything went to the textile corporation. Now the mills are not running and the property is dead. The property belongs to the private sector or to National Textile Mills. There was a move by private architects like (Charles) Correa to develop the mill lands into a single redevelopment, and they approached the (Maharashtra) Chief Minister. (They were trying to create) things needed in the city: open space, housing, and (new businesses). . . . They prepared a plan, (but it failed to look closely) at the actual ownership patterns. They made their plan and the government agreed at first that it was a good plan—justified and equitable.

Then came clashes with the scale at which land was owned or could be claimed and controlled:

Now when they said we will develop (the mill lands) like this, all the stakeholders had different views: the mill owners said, "who are you to take away a third of my land for public housing and a third for open spaces and infrastructure, leaving me with (only) a third of my land?" They went to court. The court said they are the legal owners . . . and the government was pulled from all sides.[43]

Dashmukh described disputes between mill workers, owners, and affiliates with the Correa redevelopment plan, which eventually went to the Indian Supreme Court; but in the process, and afterward, the specific policies and practices of the party then in power at the state level, the Bharatya Janata Party (BJP), framed what was actually possible in practice.[44] Fiercely opposed to the political climate of that period, Dashmukh continued,

(Those with) influence in government sent proposals and got permission. . . . Instead of fighting it out on the planning, philosophical, and land issues, what happened during the Shiv Sena Raj was that the Chief Minister himself was interested as a builder. So (often the lands) went into the hands of the builders; whoever could grab whatever lands from the mill owners . . . they tried to frame the rules. . . . (This was) not governed by *any* planning principles; the lands were simply handed over one by one without a comprehensive plan. So no housing, no infrastructure, no open spaces were built there. . . . For only five years they (the BJP) were in (power at the) state (level), but for 20 years they were in the municipal level, so everyone had an interest in land and development. . . . All the rules were changed by builders and made by builders in the absolute wrong way. . . . So all this is so messy. Even people like us {in the planning profession} had to work hard to get to the bottom of this, (to figure out what was going on), and this has created a completely distorted mess of a city. And no one knows how to correct it. That is what we are trying to do.[45]

Regardless of one's opinion about the BJP and this particular era in Maharashtra, it was clear that urban professionals—planners, architects, and urban development advocates like Dashmukh herself—were less powerful than the administrative infrastructure and its lived-practice norms. Even planners whose names garner instant recognition, and even RSIEA admiration—Charles Correa—were ultimately powerless in the mill lands redevelopment scenario. For Dashmukh, the absence of any capacity to coordinate redevelopment coherence, or to simply advocate for infrastructure and services appropriate to a coherent redevelopment scenario, produced nothing short of "a completely distorted mess of a city."

Her assessment went even further to suggest that the kind of individualized and uneven empowerment evident in the mill lands scenario completely prevented any more democratic potential to advocate for what she considered "reasonable" redevelopment.

When I asked about the relative power of large builders, Dashmukh dismissed the suggestion that they retain significant control in the redevelopment arena. Instead, she explained, with redevelopment in the hands of an uneven subset of everyone, it was "out of the hands of anyone." And yet, everyone could theoretically engage in practices that maximized their own material gain from a development-related transaction. She explained:

> The kind of mess (that has emerged in the past twenty years) now even the builders can't help. Even the *panwalla* is demanding with an old building that is to be demolished . . . the illegal tenant who occupies it has a *goonda*. It's very easy to say that builders have become the leeches who are taking the blood from the city, but now the small fries are also sucking blood from the builders. In huge numbers. The other day the commissioner was saying every small and big citizen of Mumbai has become a blackmailer, because he is in a position to take advantage of the legal system. . . . You'll see the middle class people who are staying in buildings in Hindu Colony, not only do they want the free flats which have been given by law, they want parking and money to maintain it for next 20 years, and in addition to that they want money in huge amounts. Now you can manage that with 10–20 people. How do you manage it with 400–500 tenants? So I don't think builders are very powerful . . . it was much easier to become powerful with the industrial properties because the owners were large owners. . . . Now there are thousands of tenants . . . so the economics don't work for builders. . . . It doesn't work. Even with the black money and all it doesn't work. If I was a builder I would not touch any of the redevelopment properties. I would sit quietly. It's out of the hands of anyone . . . that's the worst thing.[46]

I asked specifically about the group in which I was most interested, architects. Dashmukh's response resonated somewhat eerily with the account to which Rao alluded in his discussion of the 1930s, noted earlier in the chapter. I told her that most architects with whom I worked at RSIEA described a fraught institutional landscape, but claimed that they themselves never engaged in practices that would be considered corrupt. In fact, they made these claims with a sense of honor, invoking them as evidence that practical reform was simply in the hands of individuals. Yet, I noted, they also often located the epicenter of "corrupt" or nefarious architectural practices at the level of the municipality. Dashmukh replied with a knowing smile:

> Yes, municipal architects. The old retired people. Every retired person becomes a consultant to a private builder because he (knows) so many things and people, he knows the clauses that can be twisted or used in a twisted form. They can also help bribe people in the right place at the right time. Most of the retired architects from the BMC . . . become consultants to the builders. They are not part of the BMC but their (contacts there) remain. "Municipal architects" is not really accurate—to be employed by the municipality is one thing, and (some) might be building architects . . . and good designers. (The key is in who assigns their own) name when applying to the municipality for permissions. So I may prepare the drawings, but then I give everything to a "municipal architect." He is the expert who will modify my plan to suit the municipal regulations. He will prepare a face to face drawing and manipulate a few dimensions here and there, and he will sign the drawings as a municipal

architect. Assuming he is a licensed architect he will sign. If anything goes wrong, he will be made responsible . . . not the designer. . . . The "municipal architect" is acting only as a middle person between the municipality and the builder. . . . So this ensures that no single architect is responsible. If somebody is to be punished for wrongdoing this guy who signs the drawing will be punished.[47]

In our last interview, we returned to the Adarsha scandal, and more generally the scope for urban development reform. Is it in any way reasonable, I asked, to regard public interest in the open and green space events as something we might take seriously for its potential to actually implement changes in Mumbai's landscape? Again, the spheres of municipal and state power framed her response: her organization had proposed clear reforms, she said, but once passed to the state level, they simply had to wait and see. She explained:

Four years back we had a committee in this office (which) . . . prepared (a proposal for) a revised MRTP act. For four years it's been lying at state level. It has to go through the state assembly because . . . the chief minister can't take a decision. . . . Everyone agrees that (the act) needs to be changed, but it needs to happen at the state level. . . . (This) has become a big block. Now there is another change which is possible: . . . to revise the MCGM act and say that we don't need to have consent from the state government to approve the development plan, and we, the municipality, will take all the responsibility of making (and) administering it. The (1992) 74th amendment says why should central government interfere with state government, and why should state interfere with local authority? Everything (planning, construction, development, environment) should (happen under) local authority, but (at the same time) no one wants (all that power) because people want to blame the other level of government when things go wrong."[48]

Reform could happen, then, but the policy reform proposals that Dashmukh's advocacy group had put forward—just as the exhibitions described above had also suggested and itemized—would have to yield to a present distribution of political power that pulled between the state and the city.

As if on cue, at the time of these interviews, the then-Municipal Commissioner Subodh Kumar announced that no building permissions would be granted unless Right to Information (RTI) petitions could evidence that they had been granted in full and with accuracy. Permissions were denied and construction was stopped throughout the city. When I asked Dashmukh what had prompted this effort to stand up to builders and finally enforce the importance of legitimate permissions, she offered a singular reply: "Adarsh made the change," she said, "and many other scams are there."[49]

. . .

This chapter began with city spectacles, each predicated on aspirations to improve Mumbai's social and environmental landscape through provisions of open and green spaces. The specific publics that assembled to be part of initiatives like Reimagining Mumbai, Open Mumbai, and Breathing Space may have differed in composition and focus, but their shared basis was an enthusiastic endorsement of expanding environmental spaces in the city. As the date for a new development plan approached, their mission was buoyed by the promise, or at least the hope, that a new city blueprint would reflect their vision of better Mumbai.

A dual narrative of failure emboldened these events, focused in different ratios on the need to rectify the city's socioeconomic asymmetries and the importance of sustaining environmental functions. Whether they mapped primarily onto poverty and inadequate shelter or the catastrophic floods, the idea of failure itself energized aspirations to imagine a reconfigured urban landscape.

Yet no public exercise in imagining the future city offered a full and nuanced account of the institutional politics and dynamics suggested by the calls for a greener, more open Mumbai. The mechanisms of change were in this sense remote from the events that declared that change to be possible. The many scales of authority and bureaucracy that held urban reconfiguration in their power stood noticeably apart from the passionate, creative arenas in which the city was reimagined and redesigned.

Cultivating civic green expertise, then, itself suffered from a certain kind of subtextual failure. Until the agents of aspiration met the agents of Dashmukh's "mess," the structural reformation needed to remake Mumbai's urban landscape would remain unconnected and unresponsive to any "good design" ideal.

With city-in-the-making spectacles in mind, and the bureaucratic landscape of power to make the city in the balance, let us sharpen the frame even further to focus on a specific case of urban open space advocacy, one curiously separate from the public spectacles described in this chapter. In the chapter to follow, I consider a moment when the cause of protecting and preserving an existing open space was mobilized to serve a highly exclusive public.

5

More than Human Nature and the Open Space Predicament

Green and open spaces figured prominently in imaginative exercises in urban redesign, and they also provided a variable, an empirical indicator for the environmental success or failure of the Mumbai of the future. Assuming even distribution and access, two conditions almost never met in practice, the logic of open space seemed to offer a comprehensive bridge between ecological improvement and social justice. It seemed logically and automatically connected to promoting biodiversity habitat, better and cleaner water systems, improved coastal resilience and drainage control, better air quality, and a host of leisure opportunities for inhabitants of a population-dense city struggling for "breathing space;" these would all enhance human well-being. Here was a model for a "truly integrated" urban environment, one left unserved by the development trajectory of Mumbai's contemporary history.

Anchoring collective hope for an improved city to an idealized vision of social and environmental vitality is neither novel nor surprising; in fact, precisely such visions feature in sustainable city thinking worldwide. The presence and absence of green and open spaces—be they parks, urban gardens, urban forests and conservation areas, or expressed as the simple percentage of city land that is not built up—is ubiquitous in conventional assessments of relative environmental vitality. However, as I will show in this chapter, those same spaces present an analytical challenge to theoretical and in-practice logics of environmental justice, integrated socio-nature, and the modes of social identity reproduction fundamental to the social life of urban open spaces. At the core of this challenge is, yet again, the idea of the environment as an integrated subject, as it is precisely the integration gesture that confounds our best attempts to simultaneously understand human and nonhuman transformation.

PRESERVATION, SCIENCE, AND EQUITY: ECOLOGY AND THE ANTI-PUBLIC

A shared element of each event described in the previous chapter, and a general characteristic of contemporary middle class assumptions about desirable, livable cities with ecological integrity was the presence of plentiful open spaces, also referred to as urban open or urban green space. Generally, these terms can encompass everything from abandoned urban lots to dense urban forest stands. Together they have come to occupy an automatic place in regional and global conversations about desirable aspects of sustainable future cities.

Indices like the ratio, distribution, and scale of available open spaces often figure prominently in environmental policy objective-setting, and together with concerns about climate resilience, water and air quality, energy provision, and circuits of waste, these have become standard elements in regional and global assessments of relative urban environmental quality and performance.[1] In turn, they are often used as indicators of overall human well-being in urban contexts.[2]

Challenges to assumptions that more open spaces beget greater social equity abound; recent social science scholarship in Indian cities like Delhi and Mumbai reminds us afresh that the lived experience of urban open spaces is more often one of social exclusion, not inclusion.[3] This literature enumerates injustices and violence that may be socially legitimized precisely *because* of the place of green spaces in global and more locally active ideologies of open space entitlement. A broad literature also shows that expectations for certain modes of sociality and "civility" often accompany the regulation of such spaces, rendering them anything but open access.[4]

Global open space narratives and metrics are also known to obfuscate the complex local social processes through which access to such spaces is socially policed, calling into question assumptions that open spaces provide automatic benefits regardless of one's prior place in various strata of social difference. The nevertheless persistent "greater common environmental good" idea, famously reinforced in the classic work that defined sustainable development for the international policy community in the nineteen eighties, *Our Common Future*, continues to enable exclusive and often unjust environmental policies in the name of, and with the intention of, promoting social equity.

For the many who deem the social inequities that can accompany urban open space creation unacceptable, but who are nevertheless committed to forging a more equitable balance between a city's material built form and its vital biophysical processes—indeed, for those who, as in the CitiSpace explanation of its Breathing Space exhibition, regard open space advocacy as anything *but* elitist—social equity and justice principles must always accompany open space creation. This may mean that so-called urban greening marks an opportunity to open new social avenues for collaboration, advocacy, and consultation among those most marginalized in society. In this way, the provision of access to outdoor, open, safe spaces for leisure activity may be regarded as one of many tools for achieving higher levels of urban social justice.

In planning and policy circles, metrics that capture the proximity of *all* urban inhabitants, not just elites, to green spaces, and metrics that ensure a specific set of qualities to *all* of those spaces, are considered useful safeguards against the kinds of greening injustices that may deprive or further marginalize disadvantaged groups. Globally, many cities ratify development plans that are guided in part by target ratios of open space to population. New York City's iconic PlaNYC Sustainable Development Plan, for example, proudly declared its intention to ensure that by its target date of 2030, "all New Yorkers will live within a ten minute walk to a park."[5] Proximity and access to green spaces then, was deemed in the New York case to be as important as the existence of green spaces in the first place. Similarly, in Mumbai, despite the nuances signaled by P.K. Das and Associates' typology of open spaces, the green space category often became shorthand for human use-focused parks, and the majority population's lack of proximity to parks punctuated conversations assessing Mumbai's relative failure to secure a green urban future.

We might draw from this discussion the observation that, in case studies of city park-making, a somewhat polar analytic offers two general clusters of thinking: one notices the very powerful contemporary purchase of opportunities to remake, reclaim, and refashion longstanding patterns of urbanization. Here, to "green" a city, and to engage in "good design," are regarded as acts that promote social empowerment and equity. The second cluster marks the exclusionary and often violent ways that such initiatives are deployed in practice, and describes the social and spatial marginalities they can recreate or newly produce. Here, the social life of urban greening seems to return forms of social exclusion that replicate or reinforce existing power asymmetries in cities.[6]

Notably missing from these two modes of thinking is attention to green or open spaces as a complex category that might contain a range of open space forms, in which each form enables or dissuades particular types and intensities of human and nonhuman "use." If different types of open space perform unique biophysical and social functions, so too will their capacity to generate more socially equitable circumstances vary. Or will they?

In this chapter, I trace a specific instance in which an open space preservation effort raised questions about the category of open space itself, in part through the involvement of an RSIEA professor undertaking precisely the kind of work that, in training, the program promoted as good design. The case follows the analytical concept of more-than-human nature to consider when and how it introduces the problem of more-than-social exclusion. If we differentiate open spaces according to their human access profiles, that is, along a continuum of heavy recreational use (a running track, for instance, or a cricket field) to more restricted scenarios of human use (like a relatively dense, closed-canopy forested area), the very question of what urban environmental exclusions do may be recast, and in the process perhaps, urban environmental justice might be rendered more expansive. At issue here is the tendency to assume that most, if not all, urban green or open space worth advocating should serve direct human leisure objectives for as broad a public as possible.

. . .

A few months after I arrived in Mumbai, I received a call from a colleague at a major international conservation organization. I have worked in various research capacities with this organization over many years, and I maintain an active interest in, and periodic involvement with, their work in Asia. The call wasn't necessarily a surprise, then, but the purpose behind it was quite unexpected.

After discussing the status of my Mumbai-based research, my colleague asked if I was familiar with what were then recent controversies surrounding the large forest and temple complex in the heart of South Mumbai, the Parsi complex known as the Doongerwadi Forest.[7] My longstanding interest in the scope and distribution of green or otherwise undeveloped spaces in Mumbai meant that I was keenly aware of this relatively large, fifty-four-acre forested area in Malabar Hill. The forest was established in the seventeenth century as a sacred grove to surround several Parsi Towers of Silence, and so it was an existing urban green space among very few in the city. Yet it was also highly unusual because it was such an old, contiguous, closed-canopy forest—in fact, the only closed-canopy contiguous forest in South Mumbai (and, save Sanjay Gandhi National Park to the north of the city, much of the rest of Mumbai as well). Located in one of the wealthiest and historically most elite parts of the city, it was even more unusual because access to it was restricted exclusively to members of the minority religious group, the Parsis, and even then, only for specific rituals. Although densely vegetated, this was not a park or leisure space; instead, it was a forest whose central purpose was to shelter and seclude the sacred Towers of Silence inside it.

In fact, at that time, the Doongerwadi had gained more recent notoriety as part of an unfolding drama across the Indian subcontient. The controversial use of a relatively new veterinary drug called diclofenac had had observable and catastrophic impacts on South Asia's population of vultures. Use of the drug in livestock, upon whose carcasses vultures naturally feed, was directly linked to the fatal poisoning and a massive die-off of the subcontinental population of gyps vultures. By 2007, the drug's toxic effects were widely believed to be the primary cause of a dramatic vulture population crash: previously robust regional populations of oriental white-backed vultures had declined to the point of near extinction, and long-billed and slender-billed vulture populations had declined by 97%.[8]

Although banned for veterinary use in 2006, diclofenac remains available for human use, and that availability ensures its persistent, widespread (if illegal) use in veterinary applications. Effective, "vulture safe" alternatives to this drug exist, but none was as inexpensive as diclofenac.

In addition to disrupting basic decomposition patterns in India's livestock, the vultures' disappearance had dramatic consequences for Mumbai's small but historically relatively elite religious group, the Parsis. For centuries, Maharashtra's vultures were the key to funerary rituals prescribed by Zoroastrian (Parsi) religious tradition. In this ritual, Parsi deceased are laid out on high, open platforms in specially constructed enclosures known as Towers of Silence (*dokhmas*). There, on the sky platform, decomposition takes place through the work of scavenging birds—vultures. As the vulture population crashed, the decomposition process that had functioned for centuries on those platforms was disrupted, and the Towers of Silence in Mumbai harbored more and more deceased whose final funerary rites were disturbed and even halted. Without the rapid decomposition ensured by the vultures, bodies laid atop the towers breaking down only very slowly. The anguish and delicacy of this matter fueled outrage inside and outside of Mumbai's Parsi community, and brought to public consciousness the religious group, their ritual, the function of the towers, and the near-extinction of India's vultures. It also brought to public consciousness the extraordinary urban forest within which this drama was unfolding.

Media coverage tended toward the sensational and the macabre, adding layers of insult and pain to an already wrenching situation in the Parsi community. To the horror of Parsis and non-Parsis alike, news articles graphically described an accumulation of dead in the towers, with no natural catalysts for total decomposition. Seeking alternatives while desperate to maintain the integrity of their life cycle ritual, the community adopted experimental measures. They activated solar collectors atop some of the towers, in hopes of artificially assisting with natural decomposition, with unsatisfactory results. Debates about whether to continue the traditional funerary practices at all ensued, and the degree to which these practices were "modern" was set against the backdrop of an increasingly dire and untenable situation in the towers. The intensity of the debates gave this socioecological story a sense of urgency and purchase far beyond Mumbai, its Parsi community, and the region. Articles in the New York Times and Harper's Magazine, among others, brought the unfolding drama to the attention of readers worldwide.[9]

For as long as the Towers had existed, so too had the Doongerwadi forest that surrounds it. Originally far larger, the contemporary fifty-four-acre closed canopy forest today represents South Mumbai's largest patch of unmanaged and therefore presumably ecologically robust green space. Indeed, it is one of very few spaces in South Mumbai that can qualify in any way as "green." But as noted earlier, this was by no means a green space in the sense of an open access park or leisure space. It was a solemn forest, accessible only to Parsis and even then under very controlled circumstances. The Doongerwadi remained a forest over time likely *because* it sheltered Parsi life cycle rituals. These were sustained in large part by much broader ecological processes, most prominently the vultures and the specific patterns of decomposition their activity enabled. One might argue that the vultures had as much to do with the preservation of this forest area as the presence of the Towers of Silence themselves, or even the strict controls on human access and use beyond visiting for purposes of ritual. The stakes of a grave situation in the vulture population thus extended directly to green space, religious ritual practice, and human identity formation itself.

On the phone, my colleague explained that her organization, which has for decades based a significant portion of its programming on efforts to revitalize endangered animal and plant populations, had taken an interest in the vulture decline in Southern India. She asked if I might consider assisting with a pro bono effort to better understand the green space at the center of the vulture issue: would I be interested in looking at the Doongerwadi forest "as a green space?" That is, could I help to assess the forest's ecological value in a way that bracketed, insofar as was possible, the volatile religious and identity politics issues that had dominated the vulture controversy, and focus instead on the fifty-four acres of forest that surrounded the Towers as an ecosystem in the conventional, exclusively biophysical sense—as a bundle of ecosystem characteristics and services?

She explained further: while conducting advocacy work to mitigate the vulture die-off, some members of the Doongerwadi forest stewardship community had contacted the organization to learn about precisely these qualities. In a sense, they wished to understand the forest "as a valuable open space," but not as a public park or a place of leisure. They wondered how it functioned "as a forest," and whether that function bestowed values on the Doongerwadi beyond those that the community already knew and affirmed. For centuries the forest had been cherished as a sacred grove and a shelter for the Towers of Silence, but the community now wondered how else it might be valued, both among Parsis and by the city at large.

To compound the tensions inside and outside the Parsi community over the loss of vultures and the dysfunction it had wrought, the Doongerwadi forest faced another sort of unprecedented pressure. Over time, and under present conditions of soaring real estate values, any patches of "undeveloped," that is, unbuilt space in South Mumbai assumed astronomical value for developers. The Doongerwadi was no exception, and its coverage in mature forest only increased its already soaring economic value.

In a city rife with real estate deals and speculative investment, this was land with an almost incalculable financial value to developers, and hence, potentially, to the Parsi community. My colleague described a community torn by how to proceed, and understandably concerned that development pressures might eventually produce a more financially stable future, but at the incalculable cost of the total demise of the forest. Those who had contacted the colleague were uninterested in selling the land regardless of its real estate value; quite to the contrary, they sought to preserve it, and in doing so work to revive the vulture population. It was crucial, then, that they understand and express alternative ways to value Doongerwadi land, and one of these was clearly its value as urban green space.[10] My colleague told me that her organization wished to help them understand the forest "as an ecosystem and a natural resource" in order to supplement its known value as a cultural site, or sacred grove, and its estimated economic value as a potential parcel of real estate.[11]

Embedded as it was in Mumbai's historical moment, when the entire cityscape was in some circles being actively reimagined, and given the central role of open space advocacy for motivating public engagement in that moment, I was both intrigued and hesitant. After all, this was a highly exclusive green space. Questions of access, proximity, and claims to use of any kind outside of religious purposes were not the focus here. To assume the task my colleague requested could easily translate into helping to keep the forest socially exclusive; to illustrate its biophysical value would hinge, of course, on the uniqueness of the forest as old, as closed canopy, and as relatively unmanaged, unmanicured, and, in the language of natural resource management, undisturbed. Did I dare play a role in shoring up arguments to keep this forest closed and exclusive? On the other hand, without a set of additional ways to understand and articulate its value, this ecologically and socially significant green space would surely succumb to the power of real estate pressures. These, in the end, could prove far more powerful in present day Mumbai than any political movement to "reimagine" the city.

A great deal of political ecology research shows how scientific discourse is often used to provide falsely "neutral" arbitration in otherwise seemingly intractable socioenvironmental disputes. This left me further weary of assuming the easy role of an outside so-called expert who could give an authoritative account of the value of this urban forest "as a forest." Furthermore, my actual expertise in forest assessment and forest science was limited—a skillset developed only as a graduate student in a handful of courses. I was hardly the specialist who might evaluate and enumerate the species composition, forest health, and specific characteristics that gave the Doongerwadi value beyond its potential as a piece of developable real estate.

And yet I was also keenly aware of the unique set of questions before me. How might we understand this forest as a socionatural green space among open, green spaces that are largely imagined as parks and leisure spaces in Mumbai? Did this represent a chance to foreground the complex biophysical interplay of various *types* of green spaces, and to amplify their role in urban socioecological vitality? In some ways, I was being asked to participate in a process of literally imagining an alternative future for the Doongerwadi, one guided by an ecological narrative that would reinforce an existing cultural one, and one that would challenge the logic that anticipates urban development as a de facto response to market forces and powerful corporate-bureaucratic interests. In other ways, I was being asked to help strengthen a case for reproducing a green space almost completely inaccessible to the vast majority of Mumbai's population. In either scenario, the forest's very existence "as a forest" was clearly at stake. I was once again reminded in an experiential way that there is never such a thing as neutral research.

The narrow band of analytical engagement with green spaces that I recounted above—the prevailing inclination to read them either as opportunities to promote social justice or as spaces inevitably implicated in new forms of social exclusion—was instantly insufficient here. Yes, the Doongerwadi was a highly exclusive forest on grounds rooted in social identity and sacred territory claims. For centuries its interior had been accessible only to select groups of Parsis, but the invitation was now extended to me, to walk the forest with the community's forest steward. I interpreted this as a significant signal of how deeply fraught with urgency this matter, and hence the forest's preservation through reinvention, had become.

Back on the phone, my immediate reply to my colleague was one of gratitude for thinking of me but also a detailed claim that I was not qualified to undertake a forest assessment. Without a baseline evaluation of the species mix, age class, and general health of the trees on the site, additional surveys of the biodiversity mix the area was supporting was not possible. But my colleague had clearly anticipated this response. Suggest a botanist, a forest scientist—suggest the team that would be needed, she quickly offered. The important thing was that I would meet the Doongerwadi forest steward and coordinate a set of formal questions that could serve as the basis of a study. This would then help the community discern the area as a forest. My involvement would get the initiative started, she explained, so that it could assume its own momentum, as determined by the Parsi community.

On learning further that the Bombay Natural History Society, India's Central Zoo Authority, the Ministry of Environment and Forests, and others were already deeply involved in large-scale efforts to reverse vulture extinction trends, my curiosity was piqued. I agreed to meet the forest steward for a walk in the Doongerwadi, nervously conscious of the fact that very few non-Parsis had ever been granted such an invitation. Clearly just by granting me entry, the remaking of this forest area had already begun; my presence in that process might allow me to suggest biodiversity as part of a conversation heretofore focused on vultures, ritual practices, human identity formation, and real estate development. Perhaps it could also add other forms of open space to a civic conversation that often conflated desirable open spaces exclusively with fully accessible parks.

· · ·

In the days between our phone conversation and my first scheduled visit to the Doongerwadi forest, I studied the vulture conservation initiative proposed by the consortium of ministries and organizations named above. Their projects followed from recommendations made in at least two official plans—the Vulture Action

Plan of the Government of India (2006) and the South Asian Vulture Recovery Plan (2004). Both of these called for a complete ban on diclofenac and the rapid establishment of vulture breeding facilities.

Since 2006, state governments have set up vulture breeding programs in three Indian states. The Central Zoo Authority also maintains vulture conservation centers at five zoos, but these are not considered effective or highly functional. With fewer than 350 vultures in captivity across all of these centers, the need for additional breeding programs was regarded as critical within the South Asian environmental policy and advocacy community. So critical, in fact, that the Bombay Natural History Society, in partnership with multiple government bodies, proposed to construct two colony aviaries in the Doongerwadi forest, each "attached" to one of the three operational Doongerwadi *dokhmas*. These would then be linked to another breeding aviary roughly sixty kilometers north of the city, in Sanjay Gandhi National Park at Borivili. This was referred to in project documents as a "Main Center."

Captive breeding and vulture conservation work at Doongerwadi would require an aviary infrastructure that included two vast, completely closed nets suspended at a height of thirty feet (at least twelve feet above the rims of the towers) and surrounding each of the operational *dokhmas*. In addition, a new complex to house staff and monitoring equipment would be constructed inside the forest. Such a vast complex for vulture rehabilitation infrastructure laid bare the extent to which, regardless of the public access profile that might feature in the forest's status as greenspace, it would be a socionatural entity. All of this would fundamentally transform the sacred grove and, clearly, involve forest access to presumably non-Parsi conservation professionals.

. . .

I agreed to meet the Doongerwadi forest steward for an introductory walk. Arranging our first meeting by phone, my host offered gracious but detailed instructions for how I should dress and conduct myself when I arrived. I was to wear loose and modest garments, and to be sure my arms and head were fully covered. I could bring a notebook, but other modes of recording or photographing what we discussed and witnessed were best left behind. I was to arrive at a side parking lot, and to wait for my host to escort me onto the grounds.

Our tour did not begin in the forest, at least according to the introductory narratives my host offered on my arrival. Instead, he pointed out and identified the complex of buildings at the entrance, devoted almost exclusively to the funerary purposes for which the Towers of Silence existed and for which the forest provided sonic and visual seclusion. We sat together while my host reviewed elements of the Zoroastrian philosophy of death and dying; he explained ideas of the soul, evil, pollution, and purification, and how the built structures that culminated in the Towers of Silence enabled the sanctity of those ideas through ritual. It was clear to

FIGURE 10. Looking outward from the edge of the Doongerwadi forest, new construction looms. *Photo by the author.*

me that for him, the forest that surrounded us was in every way a sacred grove; it was a buffer between the dense, clamorous, concrete city beyond it and the very separate ritual space that in many ways defined an entire religious community. In this sense, the forest made his idea of living and dying as a Parsi not only dignified, but possible.

Nevertheless, as the forest steward, my host was also keenly interested in walking me through the forest. We stayed far clear of the Towers, but I was warmly invited to see other aspects of the forest and complex. Eloquent in heavily British-accented English and highly educated, my host nevertheless explained that he had little background knowledge in basic forest ecology or management. Assuming the position of forest steward was something he did out of care for the history and preservation of the Doongerwadi complex, and so he described himself as an eager student of the forest. As we walked and talked through its use and management, we discussed basic processes of forest growth, regeneration, and decomposition. Many of the points were quite new to him, so our walk was also a lively question and answer session on elementary aspects of forest ecology.

As we made our way along wide but ever narrowing dirt paths, we walked through a forest of easily discernable patches. As one might expect, areas more frequently traversed in the course of funerary rites were largely covered in ornamental garden plants, with few trees that forest ecologists would identify as native. But beyond these areas of garden-style management, the patches told stories of different kinds of use, historical management attempts, and even the constant give-and-take of human community uses and non-human habitat. The stands were of different species mixes, age classes, and densities.

Early in the walk our conversation turned to forest history. It is well known that the Doongerwadi forest was originally significantly larger, and that a former forest section was developed for Parsi housing, at present-day Godrej Baug, in the 1970s. For decades afterward, high-rise development surrounding the remaining Doongerwadi influenced management strategies and forest use; my host told me that over thirty years ago a private high-rise apartment complex called the Grand Paradise was built to such a height that residents living on upper floors could see the funerary platform used for sky burial in one of the Towers. Complaints from those residents, my host told me, eventually led to the closure of the visible *dokhma*. Now, with a bit of anxiety, he pointed to multiple new high rises at various stages of construction all around the forest. "The super-rich who will live in them will surely have influence," he said, worried that they, too, would see the funerary platforms and register complaints. I recalled the proposed aviary netting, realizing that it would do more than keep vulture fledglings in; it would also provide a new layer of seclusion, as these additional high rises would challenge the seclusion capacity of the forest canopy alone. This seemed to intensify the urgency of discerning new narratives of value and scientific function for the Doongerwadi.

But the high rises to come were only part of the interconnected dynamics of the forest and urban construction. In several areas, we came upon clearings or discernibly younger, comparatively homogenous, forest stands. Each had a specific history that began with a dumping incident. Whether with or without formal permission, parts of the forest had served over many decades as secluded repositories for construction waste—plentiful in Mumbai, and often difficult to properly dispose. New to his position as Doongerwadi forest steward, my host said that the details of when and how each incident happened were unknown to him, and were in any case less important in the present than the fact that the forest management response usually involved covering over the rubble and planting it with whatever seemed like a good species choice at the time. Usually this meant a monocrop, often in an ornamental or otherwise resource-intensive vegetation type.

The mixed tropical Doongerwadi forest, then, had tucked within it a mosaic of stands, some ornamental, some of which forest ecologists would classify as "native" and some as "exotic."[12] I remarked that this patchwork would make an interesting forest assessment challenge, mindful that in general, such patches fundamentally change forest structure, function, and biodiversity. To understand this specific mosaic, we clearly needed a forest ecologist or botanist.

Amid his urgent stories of forest threats, construction and development pressures, and the denuding effects of debris dumping, my host pointed to the elements of the forest he so clearly loved. Peafowl were all around us; he spoke of the existence of butterfly and plant species here that he was sure had otherwise disappeared from the Western Ghats. "We have them," he said with a mix of pride and admiration, "we just need an expert to verify their presence."

But as my host pointed to the forest elements that in his mind evidenced ill health, I realized that he and others charged with forest management also had some misconceptions about general forest ecology. He noted several dead tree branches and snags, for instance, asking if they would harm the forest and should be removed. When I explained that undiseased snags normally provide important habitat for birds, insects, and other animals, we realized together that these were potential assets, if a goal of forest management was to encourage biodiversity. Likewise, decomposing treefall that my host assumed should be removed because it might "spread disease" served as an opportunity to talk about how forest decomposition replenishes soil nutrients and ensures soil heath, and the general benefits of *in situ* decomposition in forests. Creepers in the forest canopy—which he had previously assumed to be harmful parasites—might, I suggested, be benign and even beneficial. This was true as well for a termite mound we observed, and for a range of fungi growing on trees throughout Doongerwadi.

At the close of our walk, we agreed that a useful next step would be to identify a qualified botanist or forest scientist who could conduct a formal species inventory and forest health assessment. Like the proposed vulture breeding aviary, and even my own presence, such an assessment would require the community to give further permission to non-Parsis to access the Doongerwadi. This was granted, and the senior botanist and forest ecologist from the University of Mumbai who undertook the work was also a longstanding member of the affiliated faculty of Rachana Sansad Institute of Environmental Architecture. Although the Doongerwadi remained an extremely exclusive and unusual open space, the development and open space pressures of Mumbai's present brought even this otherwise closed forest into direct contact with RSIEA.

My core interest in considering the Doongerwadi encounter here is to return to the puzzle with which I opened the chapter. As popularly conceptualized, green or open space advocacy in Mumbai overwhelmingly assumed urban park-making to be its core task, in part because of a direct association between parks and social equity. It rarely addressed the possible desirability of a citywide range of vegetation types and coverage, or the different access, use, and socioecological profiles that each might enable. Yet when urban greening advocates sought to promote ideas for a more sustainable Mumbai, the attributes of that more desirable city often depended on the vitality of various ecosystem processes and functions. Many of these processes and functions occur only in vegetated areas with, for example, permeable surfaces or contiguous land cover patches sufficient to provide food, water, shelter, and space for non-humans.

Stated differently, in open spaces in which the human use scenario is relatively light or non-existent, such as in the fifty-four-acre closed canopy urban forest that is the Doongerwadi, certain biophysical processes are underway that may depend, in part, on the very fact of restricted human use. To advocate for the benefits those processes bestowed positioned the advocate in some ways against the equal access logics that underpin environmental justice assumptions. To complicate matters, Doongerwadi's existing exclusive use profile was predicated on a narrow, identity-based logic of exclusive access—a blatant and often pernicious form of social exclusion. The calculus of social and more-than-human loss and gain at work here defied the simpler metrics that often guided pervasive open space discourses.

Yet it also cast the Open Mumbai challenge, *"How Would You Remake the City?"* in an experiential light, challenging whether a comprehensive inventory of city green spaces and socio*ecological* access profiles,could exist at all without creating or reinforcing social forms of marginality and exclusion.

Converging as it did with the peculiar bureaucratic temporality of a looming urban development plan deadline and the anxieties that attend impending, irreparable cultural and environmental loss, the Doongerwadi case thus sat at the unanticipated center of unresolved questions of what, precisely, the socionatural elements of green and open spaces *do,* and what they might be *intended* to do, in the Mumbai of the future. Was social exclusion permissible if part of its organizing logic was open space preservation or creation? Could social exclusion be regarded as an acceptable cost of maintaining a nonhuman species on the very edge of extinction? Such questions have long histories in non-city spaces, of course; could they be reasonably posed in a city like Mumbai?

One way to engage this question is to note again that the very fact of my own presence in the Doongerwadi forest was the result of a strategy for discerning and re-narrating the multiple forms of value present there, critical aspects of a defensive stance against the dual threat posed by disappearing vultures and the forest's swelling real estate value. The speculative appeal of this land suggested the real dual loss: of the vultures, and with them, a centuries old funerary ritual at the center of religious identity production. Those who wished to maintain Parsi identity through the specific funerary ritual that the forest, vultures, and towers enabled were brought face to face with the extent to which the forest itself, through its provision of nonhuman habitat, had enabled the reproduction of that identity for centuries.

Yet the Doongerwadi's particular "save the forest" agenda seemed unable to claim a comfortable place in Mumbai's active open space politics, despite its potential to prevent building development in an otherwise vegetated area. Its necessarily highly exclusive access profile, combined with the specific cultural rituals to which its very existence was anchored, disqualified it from an imagined pool of potentially shared, accessible open spaces—the mosaic of green that comprised a more "open" Mumbai. The very open space category so active in the previous chapter not only presumed a generally homogenous quality to open space itself, but also to the broader suite of uses a given open space might serve. These uses were primarily "social" in the sense of human leisure and access. They were not social in the sense of explicit acknowledgement that social and cultural practices in specific urban open spaces might themselves delineate which human groups were entitled to access it, and which were necessarily excluded to preserve it.

If not immediately compatible with prevailing open space advocacy discourses, what place might the Doongerwadi assume in responsible visions of a reimagined Mumbai? After all, one of the forest steward's central goals was to better understand the range of ecological functions and forms of life this green space helped to enable. Was there truly a space for more-than-human nature in a future Mumbai that also aspired to ideals of equity, social justice, and "open" access?

In an era of climate change and popular demand for more sustainable cities, a standard answer to such a question is often expressed in terms of the biophysical ecosystem functions that can be quantitatively expressed as ecosystem services.[13] Such services include, but are not limited to, energy and carbon dioxide conservation, often quantified as sequestration value; air quality improvement and maintenance; urban hydrology regulation (reducing the rate and volume of storm water runoff, mitigating flooding damage, reducing storm water treatment costs, and enhancing water quality); ecological stabilization through the provision of wildlife habitat, soil conservation, and biodiversity enhancement; and noise reduction. But these benefits accrue in different ways, in part depending on the characteristics of each urban green or open space. Ecologists often differentiate between them by assessing the level of *disturbance,* that is, the scale and frequency of human use, or other types of active management.[14] Examples of such interventions include, but are not limited to, clearance of forest areas for leisure activity, use of forest areas for waste disposal (even if the waste is organic), removal of downed trees, ornamental pruning and other forms of aesthetic vegetation removal, and planting vegetation species with particular management objectives in mind. Chemical inputs like fertilizers and pesticides that might be commonly applied in more park-like settings with human heavy access and use scenarios constitute another significant intervention. Since very little management activity occurs in the Doongerwadi forest, in its urban context it is not just a rare, closed canopy, contiguous fifty-four-acre forest. It is also a *low-disturbance* urban forest. But the very vocabulary that considers human access and use as a disturbance sits uncomfortably at best within conventional frameworks of environmental justice.[15]

Popular advocacy for open space in Mumbai made noticeably little conceptual space for the desirability of low-disturbance urban forest; for one thing, the term *disturbance* itself seemed to underline the problem with exclusive access, since only certain users would qualify as "disturbers," while others would not. In the Doongerwadi, sacred grove status enabled some amount of cultural and identity justification for this, but even here its power was limited to members of the minority Parsi community or those who automatically valued biodiversity as a thing in itself, so defined and so coined.[16]

But if we take seriously the conceptual transformation from social exclusion to something that approaches more-than-human exclusion, we face an uneasy

predicament: how can we advocate ecological resilience and social justice simultaneously? To foreground more-than-human-exclusion in the Doongerwadi case is
in some ways to position a low-disturbance urban forest as espousing more value
than its so-called high-disturbance, i.e., many kind of parks and counterparts. Low-
disturbance forests usually contain greater biological complexity and redundancy of
forest functions, which usually leads to more and more functional ecosystem services,
and resilience to a multitude of natural and anthropogenic stresses. It is also more
likely that plant and animal species that ecologists would classify as native, along
with greater general biodiversity, will be found in low-disturbance forests. Finally, a
low-disturbance urban forest is potentially so valuable in ecological terms because of
its structure: a mature, low-disturbance forest typically has a vertical forest structure
that is stratified into dominants, canopy, sub-canopy, shrub, and groundstory layers.
In each of these layers, vital and often interlocking ecosystem functions occur.

In short, if an expansive conceptualization of justice seeks simultaneous and
legitimate places for human and more-than-human nature, the Doongerwadi forest could be framed as far more valuable as a closed access, minimally managed
forest than any park that currently exists in the city, or, for that matter, that featured in the public exercises of "imagining" a desirable future Mumbai. Since ecologists often argue that broad ecosystem benefits, like a robust hydrological system
that ensures clean and plentiful water, or vegetation that ensures clean air, accrue
to everyone, they might also argue that a forest like Doongerwadi is extremely
valuable to everyone in Mumbai whether or not they ever set foot in it, or even
pass by. Indeed, the two characteristics—accessibility for leisure uses and optimal
ecosystem service provision—often occupy two ends of the "open space" spectrum
that this case brings to light.

A further obvious difficulty in the present case is that using its low disturbance/
ecosystem services value as the basis for saving the Doongerwadi from real estate
development would likely result in a massive, and yet sanctioned as science, disturbance: the construction of an aviary intended to save the vultures and, just as in
the case of the forest itself, the human life cycle ritual in which the vultures were
indispensable. It was unclear at this writing whether and how the ecosystem services profile of the forest would change if the aviary were to be established.

. . .

To precisely define the ecosystem services and value of the Doongerwadi forest
in accordance with the forest steward's request, a sense of the forest composition,
structure, and dynamics was necessary. This involved completing a standard set of
silvicultural assessments, typically called a forest inventory. The basic mapping of
the types and abundance of flora and fauna species, as well as a maturity matrix (an
assessment of which species are present in which age classes), were completed, all
by the senior professor of botany and regular visiting lecturer at RSIEA mentioned
above. It is here, perhaps, that we are reminded of one dimension of ecology in

practice as it intersected with the concepts and techniques espoused in training settings at RSIEA. Good design, in this case, necessitated just such an inventory; it was only in an accounting of the built and unbuilt components of the site, and a clearer accounting of the larger scales and systems to which the latter were connected, that one could claim to have laid the necessary groundwork for integrated, good design thinking. And although it was not posed as such in this case, it is precisely integrated thinking that raises the question I conclude with here: what was the place of more-than-human nature in the future of the Doogerwadi, and how did humanity stand to lose and gain in its wake?

Once the professor's reports were filed, it was possible to specify the Doongerwadi's conservation value in more precise scientific terms. Species richness, functional groups, and species traits like rarity could all be determined, and from these, certain monetary values could be derived through ecosystem services calculations. Consistent with the forest steward's wishes, a new way of knowing, and conveying, the Doongerwadi's value could now be articulated.

This returns us to the limited utility of our existing analytics when it comes to social analyses of urban greening. Recall that on one hand, a wealth of studies explore how the opportunity to "green" a city, and to engage in singular and collective exercises that rethink the relationship between humans and nonhuman nature within them, is often analyzed in terms of its potential to promote or reproduce notions of civility and citizenship, empowerment, political praxis, and the desirable place and form of "nature" in the city. Another cluster of work outlines the exclusionary and often violent ways that urban greening is operationalized, pointing to the marginalities that park-making can recreate or newly produce.

We note in this case that it is important to avoid the analytical conflation of parks and ecology, even when this conflation is restated as an ethnographic fact. The relationship between city green spaces and the ecological systems in which they nest—in the present case, for example, forest ecology—suggests an analytical imperative to remain attuned to the extent to which green space politics and their agents do or do not promote a continuum of human use and management scenarios intended to maximize the attributes that, for lack of a more robust analytic, I have indexed here through the rubric of ecosystem services. Yet the life and death stakes of the Doongerwadi case are simultaneously biophysical and social, and we are poised to miss these dimensions if we undertake a more typical assessment of the ratio of open spaces to a city's human population.

Similarly, when political ecologists describe the multiple "ways of knowing nature" that characterize any contest over the environment,[17] we have historically demonstrated the ways that science has claimed relative power and dominance, leaving us compelled to amplify other social experiences and narrations of nature as equally situated and legitimate "alternatives." But the present case forces a careful assessment of that stance; the Doongerwadi forest steward undertook a scientific assessment as just one in a range of strategies for maintaining the forest, its related non-human species, and in so doing reviving the religious rituals performed therein. The alternative bundle of values that the forest assessment was intended to enable did not constitute an overt call to elevate the scientific characterization of the forest over other formulations, or to bring back long discredited modes of conservation that sought to establish clear boundaries between territories for (nonhuman) nature and places for people. But it was—like the forest steward's own gesture of welcome to me—a layered attempt to develop and operationalize a fuller knowledge and experience of more-than-human urban nature, in part in service of mutually reproduced human and nonhuman lifeways.

In a now classic piece, the political ecologist Nick Heynen addressed a quandary similar to the one the Doongerwadi case compels us to consider in another guise: noting what he called "the production of injustice" in urban forestry, he argued that while environmental justice movements tended to demand universal and equal access to the benefits of the urban forest, "the resulting distributed urban forest may not address global environmental concerns as effectively as the larger forest islands that have resulted from uneven development."[18] The vultures, rituals, and towers of the Doongerwadi show how Heynen's quandary, which he framed in terms of global and local is, in fact, also evident at the city scale, and would certainly characterize the city-countryside continuum. The problem of demand for "equal access" is also one of nature-making itself; it risks leaving out a wealth of species, habitats, and systems on which the humans who seek access to those spaces may fundamentally depend.

If integrative, good design thinking compelled agents of environmental architecture and other urban environmental professions to promote optimal ecosystem vitality, then the Doongerwadi case also raises questions of how, when, and why differently positioned social actors deem it appropriate to create and maintain green spaces that are largely socially *exclusive*. The challenge, quite simply, is to extend the social justice imperative of inclusion and equity beyond human beings themselves. This challenge is, after all, inseparable from those presented by unequal social power relations, ideologies of belonging, and human social identity construction, but it is also inseparable from the logics through which human social groups select which aspects of non-human nature thrive, and which meet their demise. As the fate of the vultures reminds us, the two are inevitably and inexorably interconnected. The environment, as Dr. Joshi explained earlier in the book, is indeed an "integrated subject."

If the agents of RSIEA, in this case the professor who undertook the Doongerwadi forest assessment, reconnect contests over Mumbai's future open spaces to the active making and dissemination of ideas and practices of good deisgn at RSIEA, then the politics that good design thinking constructed warrant more careful attention. We return, then, in the next chapter, to the making of a collective moral ecology through the curricular experience at Rachana Sansad Institute for Environmental Architecture.

6

Consciousness and Indian-ness

Making Design "Good"

"That which is working against sustainable design is the consciousness of de-signers. Sorry if this is moralistic, but it's my observation . . . those who succeed have this consciousness through which you feel a need for holistic design."

(DR. C.L. GUPTA, SOLAR ENERGY UNIT, SRI AUROBINDO ASHRAM/ AUROVILLE)[1]

"We don't have to look outward; we have to look inward, toward our own history. We invented environmental architecture! Look at the Vastu Shastras. . . . But you won't find that (written) in LEED standards."

(DR. C.L. GUPTA, SOLAR ENERGY UNIT, SRI AUROBINDO ASHRAM/ AUROVILLE)[2]

This chapter explores moments when RSIEA training invoked specific claims about historical lifeways, categorized these as "Indian," and used them to further explicate ideas and techniques of "good design." Focusing on study tours and des-tinations, I note how a specific construction of Indian-ness was generated in the process of further explicating RSIEA's notion of good design.

With every training cycle, the suite of field visits featured in RSIEA's curriculum completely changed, with the one exception being the visit to Auroville. In the particular semesters I draw from below, our destinations included several sites in Bangaluru, Auroville, Chennai, and an "eco-village" north of Mumbai. Since specific field study destinations changed from semester to semester, however, my primary aim is not to provide an exhaustive critique of the sites themselves, in part because doing so risks a somewhat overdetermined attribution of impor-tance to them. Instead, I wish to identify and better understand specific moments when a design idea or physical feature of one of these sites was used to convey a specific dimension of the RSIEA concept of good design. I then show how these ideas and features were used to construct a specific category of "Indian-ness," suf-ficiently expansive to provide a place for a RSIEA group populated with students and faculty of Muslim, Hindu, Jain, Parsi, Christian, Jewish, and other origins. In a manner quite peculiar for a political era punctuated by a powerful politics of Hindu nationalism, I show that in forging an idea of "Indian-ness" as part of good design, the study tours reinforced a pervasive notion that environmental concerns, when understood as universals, not only transcend existing social and political disparities, but enable social categories capable of neutralizing otherwise volatile forms of social difference. In other words, by joining good design to a notion of Indian-ness, RSIEA's version of environmental architecture activated a particular kind of environmental affinity—an instance in which shared environmental imaginaries enable social collectives, solidarities, and accepted universals that find their basis in a shared idea of a common environment.[3] In a moment in Indian history marked by deeply and often dangerously anti-secular movements, mobilizing a notion that environmental concerns may not only be socially unifying but also potentially secularizing warrants close attention.[4]

At the scale of the region, then, Indian-ness in this environmental context was constructed as unifying, but in this chapter it is equally important to note that at the scale of global environmental discourse, the converse was true: the construction of Indian-ness associated with good design also provided a counterpoint to "Western" values, concepts, and practices of sustainability. That tension, and its experiential production, marks the central focus of this chapter.

In the curated experience of each site visit, faculty and program leaders narrated a version of Indian design history that "knew" distinctive modes of sustainability. Although the environmental conditions in the India of the present may be unprecedented, the consequent message was that their remedies could find resonance with, or may even be drawn from, certain environmental sensitivities that were evident in historical design concepts and practices.

A wealth of existing scholarship has critiqued the long history of discursive linkages between ideas of broadly-construed good design and notions of "Indian" history and identity. S. Paniker (2008), for example, describes the discursive florescence linking "wise" architectural design and narratives of Indian history that emerged in the wake of Indira Gandhi's assassination and ensuing political unrest. Among others, Lang et al. (1997) claimed in *Architecture and Independence* that this period witnessed a "marked shift in the architectural context, toward more traditional (Vedic and Shastraic) and vernacular ways of building which were being re-evaluated by both users and professionals as capable of offering potentially more pragmatic solutions to the perennial problems of housing and climate in India.[5,6] In this chapter, I aim to better understand how a contemporary experiential and pedagogical attempt to explicate good design and Indian-ness in environmental architecture formed the basis for a RSIEA environmental affinity.

. . .

As is typically the case in many kinds of architecture degree programs in India, field trips form a vital cornerstone of RSIEA's two-year postgraduate certification program in Environmental Architecture. Students are offered organized study itineraries to destinations outside of Mumbai at relatively affordable rates, and they are strongly encouraged, though not required, to attend. The trips often introduce a given semester, and are timed to give a conceptual and experiential foundation for the technical training offered in the classroom. They also attempt to produce a less tangible, but nevertheless important, sense of solidarity and belonging among students. This in turn ensures that the cooperative, team-based projects and assignments students regularly undertake may be completed effectively. While the study trips included in the curriculum over the course of this research included some city destinations, most were non-city sites. This gave the tours the added appeal of opportunities to "escape the city" and experience "fuller" versions of non-city nature while studying environmental architecture.

By far the most popular among RSIEA students, and most regularly offered, tour is to the aforementioned experimental city and intentional community associated with the Sri Aurobindo Ashram in Tamil Nadu, Auroville. Not far from the ashram in Pondicherry, this city of roughly two thousand has explicitly aspired, since its founding in 1968 by the followers of the Indian spiritual leaders Sri Aurobindo and Mira Richard (known more commonly as the Mother), to become, as the city professes and its residents repeated to us, "the city the earth needs." Consider this RSIEA student's description of her personal anticipation of the "Toward Sustainable Habitats" study tour in Auroville:

> Auroville has had, well, a certain "aura" about it. As a student of architecture, I had been hearing about Auroville for a number of years but never had the opportunity to visit it. I had heard from a number of friends and colleagues, who had visited the place to attend workshops, about what a fabulous place it was, but was unable to comprehend it completely not having had any first-hand experience myself. That finally changed when we were taken to Auroville for the workshop "Towards Sustainable Habitats," being conducted by the Centre for Scientific Research, as part of our M. Arch course. The topic or subject matter for the workshop itself was so intriguing; I found myself looking forward to the workshop even more. I think it has something to do with having lived in Mumbai for most of my life and as such, never having had the opportunity to experience anything other than the crowd, noise, and the concrete jungle that is this city. I was looking forward to experiencing another way of life, and I was not disappointed.[7]

One of the groups I accompanied to Auroville travelled to Chennai via rail or air, and then a shared bus from Chennai, with additional brief stops en route. On this trip, stops included two sites typically used by school groups of many ages for narrating "Indian" history and vernacular forms, Dakshinachitra and Mahabalipuram.

Before describing the Dakshinachitra / Mahabalipuram / Auroville trip in more detail, I note here the second study tour that I will recount later. Like the

Auroville tour, a study tour to Govardhan Ashram and Eco-village occurred near the beginning of the new semester, in March of 2012. The two day, three night trip involved a bus journey to the 60-acre ashram and eco-village site at Galtar, about 100 km north of Mumbai and located in the Sahayadri Mountains. Among other attributes, this area enjoys a "biodiversity hotspot" designation from the International Union for the Conservation of Nature. Like Auroville, Govardhan has explicit links to an internationally recognized and organized spiritual practice and philosophy. In the words of its own promotional material, the ashram is "a project dedicated to His Divine Grace A.C. Bhaktivedanta Swami Prabhupada, Founder and Archarya of the International Society for Krishna Consciousness (ISKON) and inspired by Radhanath Swami." Govardhan is just one of the more than five hundred ISKON-affiliated temples, ashrams, centers, schools, and restaurants across the world.

Yet unlike the oft-traveled destination of Auroville, Govardhan was a relatively young initiative, established only in 2003 and still in stages of construction in 2012. The RSIEA trip was experimental; ours was the first tour RSIEA made, and so faculty in particular were not only seeking to use it for teaching, but also learning for themselves whether it was an appropriate field study site. Govardhan did not enjoy the same anticipatory mystique that students attributed to Auroville, but the promise of a self-professed "eco-village" made the journey appealing nevertheless.

. . .

The social and pedagogical process of linking specific study site attributes to good design and Indian-ness often hinged on overt or implied narrations of spirituality and "consciousness." While RSIEA faculty and students rarely invoked specific religious texts to explain good design, they did make repeated references to "spiritual practice" and "tradition."

In Auroville, the city's very existence is predicated on adherence to the spiritual interpretations, teachings, and philosophy of Sri Aurobindo and his primary follower, Mira Richard ("The Mother"). Sri Aurobindo famously reinterpreted a range of Vedic texts; his "elaborations" or "revisions," depending on one's point of view, led him to develop his philosophy of Integral Yoga. To follow Aurobindo, then, is to adopt Integral Yoga as foundational to correct and appropriate spiritual engagement in the contemporary world. Despite this, our field visit to Auroville never detailed, or even mentioned, Aurobindo's philosophy. While there was a brief orientation to the founders of the ashram and to the city, our Auroville program was dominated by an agenda framed as its title implied: Toward Sustainable Habitats. A less overt but nevertheless omnipresent sense of reverence for something repeatedly referred to as Indian "history" or "wisdom" infused the Auroville-based environmental architecture learning experience.

By contrast, our hosts at Govardhan Ashram and Eco-village professed an overt commitment to what they called the "Vedic lifestyle;" the site represents itself publically, in fact, as a living demonstration thereof. Consider the following excerpt from the organization's promotional literature:

> Govardhan Eco-village illustrates "Simple Living & High Thinking"—a principle which is so succinct, yet profound, and formed the basis of life in the bygone age of wisdom. Life in the Vedic times was focused on Service, but not on exploitation; this was the cardinal rule of living and the very essence of people's dealings—with each other and that with Mother Nature. With the concepts of eco living being innate, the Vedic lifestyle was truly an eco friendly way of living life as instanced in the timeless Vedic scriptures like SrimadBhagavatam and Bhagavad-Gita. We at Govardhan Eco-village hope to present this model to the world as an alternative way of lifestyle and perhaps a solution to the impending ecological crisis. . . . The purpose behind Govardhan Eco-village is twofold—one is to present a sustainable living model based on community living and second is to educate people in the field of traditional sciences including Yoga and spirituality. . . . Since its inception in the year 2003, Govardhan Eco-village has made steady progress in Organic farming, Cow protection, Education, Rural development, Alternative energy, Eco friendly constructions and Sustainable living. In the scenario where environmental crisis is on the rise, Govardhan Eco-village is an example of living in harmony with nature.[8]

Here, certain modes of relating to environmental processes and resources are termed the "traditional sciences;" these are then undertaken as demonstrative of the ashram community's commitment to the "Vedic basis" for sustainable architecture and, in fact, all aspects of sustainability's moral parameters in social life.[9]

Study tours to these sites inevitably focused on the architectural practices and features to be observed in each, but our hosts' narrations of the principles of good design that produced the places themselves offered a kind of contemporary evidence that the *idea* of Indian history—less as a bundle of texts or repertoire of rituals than as an enduring set of wise guidelines for environmentally responsible living—was thriving in each site's material form. This precolonial historical "basis" for the material and social dimensions of sustainability that students could observe in real time and form underpinned consequent claims that linked good-design in environmental architecture to a specific construction of "Indian" identity.

This does not, of course, mean that faculty and students automatically and uncritically accepted those claims. Full or even partial acceptance was never preconfigured, or complete. Nevertheless, the socially inclusive and simultaneously spiritual and secularizing dimensions of historical narratives of Indian good design gave it a particular appeal.

. . .

Once assembled in Chennai, our bus filled with eighteen architect-students, faculty, and one visiting anthropologist made its way toward Auroville. En route, we made intermediary stops, the first of which was at Dakshinachitra . A site described in its own promotional literature as "a center for the living traditions of art, folk performing arts, and architecture of India," Dakshinachitra opened in 1996 as a project of a non-profit organization called the Madras Craft Foundation. The complex was designed by the architect Laurie Baker and is widely visited by students of architecture and other interests alike.

Our stop there was unstructured, so there was no singularly narrated experience of the place. Students moved in small, self-selecting groups through a landscape of what the site's promotional literature calls "heritage houses," each labeled and organized along streetscapes modeled after Southern Indian regional vernacular architectural styles. In all, there were seventeen structures to explore, and the walk between them was an experiential sampling of specific and highly stylized representations of what were referred to as "typical" or "authentic" South Indian vernacular architectural forms. Explanatory plaques associated each structure with specific southern regions and identity groups.

Moving between different clusters of students and faculty, I walked from built form to built form, experiencing the physicality of carefully rendered re-creations with names like, "Kerala House" and "Syrian Christian House." One environmental architecture student wrote, in a post-trip reflection: "(this place) had the magnificent character and style of Kerala, Tamilnadu, Andhra Pradhesh and Karnataka. It (had) architectural details and elements which you never get to see in one place. It was a place where you can find all types of traditional architecture of South India."[10]

In the absence of a scripted tour or single guide, our movement across this spectacle of historical architectural styles prompted rich conversation. Students and faculty noted certain attributes for their aesthetic qualities and commented on the extent to which certain building features promoted thermal comfort, natural lighting, practical uses of regional materials, and other aspects of good design.

Though interpretive plaques marked each building as indicative of an historical place and identity group, there were no references to the social hierarchy and power relations that would have produced vernacular spaces in any specific moment in history. Offering no context for a given building's historical maker or dweller, to say nothing of their social positionality, had the effect of bundling all of the structures together as a set of politically neutral, regionally representative examples of good design, each at risk of becoming, or already, lost or unfamiliar forms in the postcolonial modern present. The buildings were meticulously constructed and presented as both exhibits and as forms to be traversed, explored, and experienced. Moving between clusters of students, I was drawn into conversations about the aesthetic contrast between the vernacular homes and the contemporary, modern building forms we had just left behind in Chennai.

With Dakshinachitra the first stop on the tour, we began by physically moving through a curated cluster of regional vernacular forms that stood for a certain dimension of the regional past, seeking in them good design techniques for the present. At the same time, we were locating forms and ideas through which to trace the regional origins of "Indian" sustainability.

The student reflection noted earlier went on to observe:

> Dakshinachitra, which is a craft centre and exhibition space all rolled in one, was a fabulous campus. . . . Stopping here, I felt that our workshop on "Sustainable Habitats" had already commenced because these homes were prototypes of the kind of life that was till we became industrialised and consumerist.[11]

A historical sentimentality thus joined with the aesthetics of the forms, leaving questions about the selectivity, stylization, or power relations embedded in those same forms of little expressed interest to most students. The point, it seemed, was not to critique their historical context, but rather to seek in the forms a set of qualities that could be regarded as enduring, and therefore timeless; here was a first hint of attributing a socially neutralizing power to this particular idea of sustainability.

Rao (2013) reminds us that such a quest to recover "harmonious" architectural technique from a vernacular past almost erased in the industrial present is not new. For him, it is a 1939 essay that epitomizes this for the case of Mumbai. Titled "Traditional Domestic Architecture of Bombay," Rao writes that the author, Janardan Shastri, "appeared to sense that something harmonious was being irrevocably destroyed by planned, regulated building of the sort undertaken by certified professional architects."[12] The essay gives a historical account of Bombay and relates different social groups to different types of dwellings. In it,

> He selected houses he considered "typical" of some of the older neighborhoods of the city, such as the Fort, Kalbadevi, Girgaum, Parel, and Mahim. The mediating link between the people and their dwellings was, for Shastri, religion or dharma. Hindu life was so saturated with the notion of dharma, a concept that cannot be abstracted into a category like "religion," that it also suffused the Hindu dwelling.[13]

This in turn created, according to Shastri, aesthetic continuity. The coming of the Portuguese, and eventually the British, brought changes that Shastri uses to explain the break that ensued; there is deep nostalgia in this piece "for a time when buildings were authentically Indian."[14] A major force in the disruption of an imagined precolonial harmony is the hybridized figure of the builder, contractor, and developer: "In the absence of an overarching Hindu cosmology, redemption from the godless, profit-seeking purgatory that is the modern city is only offered in the synthetic vision of the architect."[15]

Another series in which Rao takes interest appears in the *Journal of the Indian Institute of Architects* called "Lesser Architecture of Bombay." These pieces, which appeared in the 1930s, "represented an acknowledgment of the widening understanding of what constituted architecture in the Indian context. "Most importantly," he argues, "this series adumbrates a compromise between "traditional" and "modern" dwellings."[16]

It is useful here to underscore the malleability of terms like "vernacular" and "traditional," particularly in the instances of environmental learning that took place on RSIEA field visits. Tamara Sears (2001) provides a useful distinction between these terms when she calls vernacular "something that grows out of lived experience, that is embedded in the social, cultural, and environmental conditions in which people conduct daily activities, and is therefore intuitive to the people who produce it."[17] Vernacular architecture, then, can signal the "informal, usually domestic architecture that is rooted in local tradition and is generally produced by craftsmen with little or no formal academic training."[18] Sears distinguishes this from "tradition" as "a larger category, encompassing a variety of assimilated phenomena, of which "vernacular" becomes one part."[19] Distinguished in this way, we notice that the vernacular is dynamic, so moments when it is "fixed" in time and reproduced as valuable, such as in the structures we toured at Dakshinachitra—and the specific contents, materials, styles, and forms that were implicated as a result—are important for understanding the construction of "Indian-ness" that informed the RSIEA concept of good design. Sears continues:

> In the discourse of colonialism, vernacular was seen as something inherently indigenous, pre-modern, and uninformed by the concept of civilized society emerging from the enlightenment. In more recent times, vernacular and traditional architecture has become a glorified notion, and, as Miki Desai noted, an aristocratic folk paradigm often emerges when scholars and practitioners talk about tradition. The authenticity of the vernacular is often praised, and educated scholars, preservationists and other elites often seek to save what they see as a tradition going extinct in the face of the onslaught of rapid changes brought on by modernity.[20]

Clear tensions emerge here, as the agenda at Dakshinachitra was to learn good design techniques from a historical vernacular: as we noted points of evidence of good design, the puzzle was whether and how we might import those elements from the past to the present was quite obvious, as was the impossibility of making so many of those same forms relevant in the population-dense urban context of Mumbai itself. Perhaps in this way, our visit to Dakshinachitra was less an encounter with the vernacular for its demonstration of " assimilation, change and hybridity (as) ongoing processes," and more of a gesture of "sifting through (the vernacular) to find the truly pure."[21] At least, it seemed, the truly "good" in good design.

. . .

After a few hours touring Dakshinachitra, the group returned to our bus and journeyed further to the coastal town of Mahabalipuram. A significant seaport as early as the first century, Mahabalipuram is historically associated with the Pallava Dynasty and is comprised of dozens of carved rock structures that remain essentially intact. These include magnificent rock-cut temples, sculptures and reliefs.

The site was a remarkable spectacle of resilient built forms, made even more powerful by the harsh coastal setting. A coastal location may signal automatic vulnerability in an era of climate change, but Mahabalipuram seemed to embody resilience—its structures still standing, and remarkably intact, after centuries of coastal weather and change.

Our group traversed this site with a palpable air of reverence. In contrast to Dakshinachitra, this was a cluster of built forms that needed no reconstruction or re-enactment. These forms, a RSIEA faculty member reminded the group, were widely considered to mark the material beginnings of Dravidian architecture; before us were strking shrines to Shiva, Vishnu, and other Hindu deities that had withstood centuries in a harsh coastal climate. Nearly all of the RSIEA students spoke of learning of Mahabalipuram in their school textbooks, and many had visited the site in past field trips. This place had iconic, national identity-making power far broader than our good design mission, but in this context it stood as evidence that good design could be traced to deeply histories.

As I walked among faculty, one noted that the diversity of forms at Mahbalipuram indicated that the area was also a center for teaching, joking that we should see it as a kind of analogue to RSIEA. "Their carving skills are our environmental architecture skills," she said. The point was not lost in the laughter; we were students on what may have been a site of ancient learning, seeing before us a striking example of efficient, undeniably resilient use of local materials. As we boarded the bus, Dr. Joshi joked with a student, saying, "See again! Sustainability is just ancient common sense!"

I note again that no mention was made of how this site was invoked, or why, in other teaching contexts, to say nothing of critiquing its use in constructing narratives of Indian national identity. Our collective purpose seemed focused on the structures and their resilience, not the social or political dimensions of the past they emerged in or the present that glorified them. To do so, after all, risked attaching this place to certain social identities or foregrounding power relations that, in the good design sociality of RSIEA, found no explicit place or endorsement. If each site simply illuminated some dimension of a multi-dimensional notion of Indian-ness, then each could provide welcome and potentially useful guidance in service of good design.

We boarded the bus yet again to complete the journey to the place that so many students were eager to experience, Auroville. Through many hours on the bus, a Tamil film, sing-alongs, and constant laughter consumed our attention. A light, almost celebratory air of excitement combined with the freedom of the journey far from home and the hours passed quickly. By the time we reached Auroville, the sun had long since set, and a deep darkness veiled our surroundings. Tired, we poured out of the bus, and into large, shared dormitory spaces in a ferro-cement building called Mitra. We set down our bedrolls, and soon fell asleep.

At the light of sunrise, we gathered around the two massive teakettles that had appeared on the landing. Within half an hour we climbed back on the bus, only to arrive just a short drive later at Auroville's Center for Scientific Research. As we disembarked, it was clear that our training at Auroville would begin here. We marched up its stairs, and into formation for the women to receive jasmine hair garlands, for all of us to pick up our information packets.

Midway through the week's itinerary was a group trip to the Matrimundir, the city's iconic spiritual center, and in many ways its distinguishing global symbol. The mundir's unmistakable spherical form, plated in gold and fused with glass, underscores the influence of Sri Aurobindo's religious philosophy on the city and its making; the unique design also makes clear that the "city the earth needs" traces its basis to a specific spiritual practice that has historically guided its approach to urban design.

The presence of the mundir on our *Toward Sustainable Habitats* itinerary suggested that this religious center also had techniques and concepts of good design to impart. Formal lectures would later refer to the Matrimundir as a symbol of "perfect human consciousness emergent from the earth, or from the old, unsustainable human consciousness."[22] This would be the conceptual frame through which Auroville's instructors would narrate the city's built forms throughout the week.

As we settled in our classroom seats, our primary instructor introduced himself by his first name, Tency. His accented English suggested Austrian or German origin, and the ashen hair swept over his neck invoked a kind of bohemian aesthetic. His voice was gentle but austere as he delivered his first lecture. It felt somewhat disjointed, then, when Tency invited us to begin our week of studying "sustainable habitats" by discussing our personal passions. My field notes from this first session read,

> (Tency) distributes name tags and asks that we identify our personal passion as he hands us our tag. My name is read first, and I find myself anxiously blurting that my passion is teaching and learning. I am immediately struck by how utterly different Tency's pedagogical style is from the classrooms back at the RSIEA, and I wonder how these architect/students will respond to his style . . . After we each somewhat uncomfortably state our passion, Tency begins a puzzling presentation. It features a slide of the Rosetta stone, slides showing other examples of carved stone tablets, and it seems to be generally about writing systems. I try to discern a clear pedagogical message, but I can't. Tency concludes this section by telling us that each day we'll begin with a short "passion presentation," and then we will listen to some music. He then plays that day's music, a video of a French fusion musician flanked by Hindustani classical musicians. The musician strums his guitar in combination with a tabla and sitar player.

As the music ended, Tency shared a series of images. A first slide read, "THE CITY THE EARTH NEEDS." He acknowledged that while this may sound like an arrogant label for Auroville, it was a simply a declaration made by The Mother, and something Aurovillians truly believed. This was the first time we heard an explicit reference to The Mother by name, but the ubiquity of her photo in the buildings we had visited made her difficult to ignore. Tency explained that those who live in

Auroville "don't like a lot of publicity," and that as an experiment in sustainable living, the city has "gone through some tough times." "All around us," he continued,
we see that we are not on a planet that is solving its problems, so perhaps this is,
despite its flaws, truly the city the earth needs." He then framed the week's course
on Sustainable Habitats as a guide to some of the steps we can take—as architects
and as human beings—to actualize that needed city. "The conditions in which we
live are a result of our state of consciousness," he said. "The external, built world
reflects only our inner state of being."

As noted in a prior chapter, Auroville was founded in 1968 by the followers of
Sri Aurobindo and Mira Richard (the Mother). Its master plan is set on 3500 acres,
arranged in a circle that is 2.5 km in diameter. Eighty percent of the land area in the
master plan is owned by Auroville, but Tency explained that real estate values in
the region were skyrocketing, and development pressures on the remaining land
that Aurovillians wished to acquire was intense. Tency offered no details of the
city's formal governmental or institutional structure, nor did we glean the city's
relationship to the Indian state of Tamil Nadu, in which it is located.[23] Instead,
Tency offered us Auroville's ecological origin story.

Nearly all of the land inside of Auroville's contemporary boundaries was, he
said, severely degraded forest when Sri Aurobindo's followers first committed to
building the city. Their initial grueling task was to re-vegetate the landscape; this
was nothing short of a massive undertaking, and it formed the basis for regarding
Auroville as a fundamentally "green" city.

Tency continued by noting the key biophysical challenges facing those who
design the city's buildingscape and conduct its urban planning. In 2012, Auroville's
population was only two thousand persons (representing forty nationalities), but
its founding mission was to grow to a city of fifty thousand. In order to do this while
remaining ecologically viable, the city would need to secure forested "greenbelts"
on its periphery, and ensure a stable and adequate water supply. Water was a particularly vexing factor, and so, Tency explained, as the city has grown so too have
experiments with urban-scale and building-scale techniques of water recycling,
wastewater treatment, and decentralized water purification. Tency also discussed
the challenges of providing adequate and ecologically viable energy supplies; he
noted Auroville's extensive use of photovoltaic technologies and energy-efficient
thermal comfort strategies. Rather than air conditioning, for instance, Auroville
buildings use in-house dehumidifiers, which allowed them to achieve at least a
quarter reduction in energy use relative to conventional air conditioners.

Tency concluded his introductory lecture by declaring that, "this is a city that
desires and expects to grow." With that, we were invited to share tea on the building's sunny terrace. When we reconvened, the session assumed contours more
familiar to a roomful of architecture students; a landscape architect introduced as
Aditya, and a town planner introduced as Lata, offered more substantive ecological

overviews of the city, and the sustainability principles that inform Auroville planning and design at multiple scales.

Aditya used an international planning and landscape design genealogy to frame his lecture. He described his primary inspirations as Ian McHarg, whom he called "the father of ecological planning" and the American author of *Design with Nature*, and Mary Jane Coulter, an early twentieth century American architect whom he called a "landscape-sensitive architect."[24] He then offered a list of thematic criteria essential to designing for sustainability at large scales. These included an analysis of the site's biophysical history, which he called its "ecohistory," a long-term trend analysis for hydrological patterns, clear articulation of the site's topography and water regime, and, in the category he called "human ecology," demographics, land use and control regimes, transport networks, and future physical projects. These combined with a basic physical assessment—climate, geology, soil types, surface and subsurface hydrology, and biological flora and fauna—to form the baseline knowledge sufficient to proceed with good design.

But we were quickly reminded of our context as Aditya turned unexpectedly to the importance of the Mother's vision for Auroville. He explained that the land development pressures outside of Auroville were unforeseen by the Mother; so too was the fact that older visions of ecological vitality which had valued greenbelts outside of cities would later be discredited and replaced by planning strategies that emphasize integrating green spaces *in* cities. "What we want to escape as we develop the city further, and as we try to influence land use policies around us, is the fate of becoming a weekend resort city," he said. "Planning is awareness raising," he continued, "because most people live life in a sleep mode."

This last statement completed a narrative arc in which Aurovillian environmental architecture was described as a product of inner consciousness, specific techniques, and intentional social action. The field examples we would visit in the following days referred back to these elements; every site visit was led by that site's creator or caretaker, and each was narrated for its value both in solving the specific biophysical challenges and for addressing the framing spiritual challenge to become the "city the earth needs."

That first afternoon, we would visit a private home called "Newlands," fully designed and built by a German architect who introduced herself as Regina. Tucked into the dense forest in a way that made it seem simultaneously integrated and distinct, the home was indeed an impressive example of ecologically integrated architecture. All of the materials, Regina told us, had come either directly from this forest or from the Auroville site. Using energy and machinery as minimally as possible, Regina had literally designed and built the home herself. The fourteen-year-old structure made extensive use of waddle and daub, particularly impressive given seasonal monsoon conditions in Auroville. As the first built form we encountered outside of a classroom setting, and as the first example of the architectural experiments that were possible in a city like Auroville, the home made a significant impression on many of the students. One reflected later:

> New Lands . . . was located in the green belt zone. Only naturally and locally available materials can be used for construction in this zone. And indeed this house belonged to this zone. With rammed mud walls and in situ seasoned wood and bamboo, this house was truly a master piece. The windows were framed with a type of seasoned and burnt bamboo known as Buddha's belly. The windows looked like a masterpiece by Gaudi. Nature was flowing in and out of the house. I was speechless.[25]

With some in the group charmed and others astonished, we moved on to Gaia Gardens, a large garden and multi-building yoga complex built and hosted by a man who introduced himself as Kireet. Here, we learned about the rainwater harvesting techniques used in the gardens, and we hiked Utility Canyon, a massive erosion-born canyon that has been a restoration focus in Auroville for decades. Kireet coordinates the financing and operation of a series of check dams intended to capture and redeposit Canyon silt, and in so doing stop large flows of run-off from further eroding the hillsides. Again, at each site, our hosts offered their personal story of inner consciousness development and its direct expression as environmental stewardship and even, in Kireet's case, environmental restoration. The students devoured the technical details of the designs, their consistent interest deepened by the repeated, passionately professed basis of these practices in "consciousness."

Very few of our instructors claimed South Asian descent, but by emphasizing that good design was fundamentally linked to "consciousness," they reinforced the idea that a unique, and longstanding, "Indian" wisdom was fundamental to the effectiveness of their work. By virtue of their diverse national and ethnic origins, they conveyed an inclusivity to the category as well; it had the communicative effect of backgrounding our differences so long as our common commitment to "sustainable habitats" through spiritual consciousness was constantly reinforced.

In the days that followed, conceptual and technical sessions further explicated good design through "sustainable habitats." Sessions like "Green Home Technologies," "Auroville's Architectural Diversity," and specialized course sequences on topics including decentralized water treatment strategies, local materials processing and use (stabilized compressed earth blocks and ferro-cement, in the case of Auroville), and renewable energy filled the schedule. We visited several other buildings, each with compelling names like, "Luminosity" or "Creativity." In every instance, those who guided us through the building described a relationship between consciousness and good design.

As our study tour was coming to a close, we reached our program at the Matrimandir. Unlike the other sites and other sessions, no host guided us through this day. The massive spherical building was not narrated for its sustainable aspects or building techniques, yet after many days of guided thinking, we were conditioned to encounter it as at very least, a source of the consciousness so central to good design.

Student reflections like these, composed after the visit, emphasized the personal experience of consciousness that many described in connection with their time in the Matrimundir:

As one walks closer and closer to the Matrimandir the mystery of the architecture begins to unfold. The paths through the petals draw us towards the centre of the globe. Everything appears to converge . . . then we start to realise the existence of each golden concave disc on the globe. As one enters the mandir, we are asked to wear the socks that are provided. This is a transition zone, where one starts to stop thinking about the worldly endeavours. Then we enter the great hall . . . a spiraling ramp. The salmon pink ambience and the volunteers dressed in white give it a feel of a sci-fi space ship. Now the thoughts are guided towards the ray of light that emerges from the top and escapes into the floor. After this transition there is the meditation hall. A huge void with columns creating an isle for seating and circulation. It is extremely difficult to realise the scale of the space due to absence of any reference. Only the vertical ray of sun entering though the ceiling and entering into a crystal can be seen. There were no thoughts. Even after trying very hard to think, there were no thoughts. Just blankness. There was silence. Silence as I had never heard before. It was a silence that I defined as the "screaming silence." Silence so intense that one could hear his own blood pressure. Silence that will make you realise . . . your own shoddy existence. The hazy white atmosphere was suddenly lit and I realised that the 15 minutes of meditation time was over. . . . The feeling after coming out was bliss.[26]

Walking into the room, nothing else existed. I walked right into the ultimate emptiness. The whiteness is both light and dark at the same time, casting sacred robes around every person. I sat, I breathed in the awe of this power, this perfection. Meditating, "om," the sound of the universe came in and out through my every breath. I at once became the perfect sphere within which I sat, and this sphere grew outward to consume all of Auroville and all of the world. We were only allowed to remain for ten minutes, which was enough for the time. Exiting, I felt endlessly rejuvenated. I had deepened so much, emptied so much. I found a sense of peace that existed infinitely inside of me. From feet to my smile. Then we all sat underneath the globe in a circle, people from all over the world, around a lotus-shaped structure. It was almost flat, but with pure white lotus petals. Where a crystal sphere sits. Water flows slowly over the petals down to the center, but so slowly that you can observe every ripple, like birds of water diving into paradise. Then exiting the Matrimandir, I awoke to the world. I walked out into nature, bringing with me in my heart and hands all that my spirit had felt—perfect cleansing. I sat beneath the banyan tree, leaning against a trunk for support, seeing no end to peace.[27]

This so-called "abode of the Mother," who along with Sri Aurobindo was and is the stated reason the city exists and persists, provided a strong symbolic spatial conclusion to a week in which the technical details of "sustainable habitats" were conveyed as only as powerful as the consciousness one brought to them. It solidified the connection between good design and a form of consciousness, even if the precise contours of that consciousness remained amorphous.

I do not wish to imply here that students received the good design-consciousness connection somehow automatically; on the contrary, each student made her own sense of the week's course content and the individual and collective experience of Auroville. Another student reflection conveys a sense of the questions left open and unanswered as the week came to a close:

> Overall the experience of Auroville has taught me many things. It showed the importance of simple living, caring for the society. It also brought about certain questions. Can we consider it as an ideal for sustainable society? Can it be replicated elsewhere and accepted by people? Or should its existence remain as something different from the mainstream society?[28]

While Auroville was in some ways a "city," its population and density suggested little that could be scaled to, and produced in, a city of the scale and density of Mumbai. Many of the innovative techniques we studied in such detail, like decentralized water management systems or mud brick manufacturing, seemed to have little immediate relevance to Mumbai and its challenges. The practice of environmental architecture, then—the ecology in practice dimension of their training—was not what was primarily derived from this study tour. Instead, the idea that good design depended in an essential way on both techniques *and* consciousness was the takeaway; the remarkably inclusive notion of Indian-ness that Aurovillians assured us derived from this duality only made the idea of Indian environmental architecture make more, and more appealing, sense.

By the close of the visit, our group showed discernible signs of an emergent environmental affinity, one that traced the general sense that "consciousness" formed an important dimension of the good design we wished to understand, as did identifying that dimension as linked to an expansive and, at least in an implicit way, a rather politically ambivalent notion of Indian-ness. We could be effective environmental architects, and viscerally understand good design, it seemed, without having to "opt in" to the messy and often opaque political economic questions and structures that determined the course of so much urban development. The essential reflexivity when it came to good design, then, was environmental, ecological, and integrated; it was not political or economic, save as an item in a site visit analysis.

Tamara Sears reminds us that, "the self-conscious and intentional assimilation of forms, for the purpose of creating new social identities, has been an ongoing phenomenon in architectural history and practice. In this sense architects have the potential to make real interventions in society as the mediators between individuals, tradition, and government policy."[29] Yet here, in Auroville, we gleaned little that might guide us to be strategic mediators, effective practitioners of good design. The aspirational life of good design was unquestionably crystalized, which emboldened us, but the sociopolitical reflexivity we would need in order to bridge that aspiration to effective ecology in practice was left, as a consequence of omission, to the domain of others. The journey, in any case, was lauded as an overwhelming success.

. . .

At the RSIEA Opening Day Welcome Program in 2012, a short program item had featured a presentation by Lucky Kulkarni, a representative of Govardhan Ashram and Eco-Village. This would be the first study tour of the semester, and unlike Auroville, it was a new and experimental addition to the RSIEA curriculum. There was no doubt that it sounded perfect for our purposes: Kulkarni promised what seemed to be an ideal eco-village setting, distributing multiple colorful brochures that promised serenity, sustainability, and spiritual growth.

Her comments were animated by colorful images of bucolic landscapes and cozy lodges. Kulkarni told the audience that Govardhan is "a village planned with urbanites in mind—the place where urban dwellers can come to experience nature." It was comfortable, she assured her audience, *and yet* natural. She continued by describing the lodging accommodation at Govardhan as "the way we (urban dwellers) want to stay, but with a rural touch."

The eco-village was designed by the environmental architect Chitra Vishwanath, and buildings throughout the complex were designed for maximum, if not complete, environmental self-sufficiency. From generous use of locally available building materials (which included compressed earthen blocks that we later learned came from Auroville), to interlinked village-scale solar energy systems, bio-gas systems, organic food production, on-site sewage treatment, small scale cottage industries, and extensive cowsheds, Kulkarni narrated an idyllic setting in which to both learn and experience the fundamental elements of good design in environmental architecture.

Unlike the much longer journey to Auroville, the drive to Govardhan from RSIEA was just three hours long, allowing students and faculty to make smaller group travel arrangements. I joined faculty members who chose to drive to the ashram the night before our two-day workshop began; some students arrived early the following morning, and a final group joined us for one intensive final day.

Our small faculty group arrived as the first from RSIEA to reach Govardhan; two young monks greeted us at the village gate and showed us to the yoga shala. In almost total silence, and with virtually no introductions, we were served a light vegetarian dinner and shown to a lodging center where our guest rooms were pre-assigned. Aesthetically, the building nearly blended with the landscape; functionally, it used a solar water heater and drew energy from sources across the property. In the rooms, a collection of Ayurvedic toiletry items produced in the village supplied us with our basic needs. What was completed seemed integrated and potentially quite interesting, but we also noticed right away that a surprising portion of the eco-village complex was still under construction.

As we retired to our rooms, we joked about the very early morning start time, and the first item on the agenda: "morning prayers and short class on mantra meditation." For a group more accustomed to a concept of "consciousness" that did not map to a specific set of prayers or even a specified religious tradition, this struck us as an unusual—if perhaps unsurprising—way to start a weekend course on environmental architecture. Agreeing we would keep an open mind, we woke according to the schedule and assembled again, some bleary-eyed, at the yoga shala.

An adult monk arrived, accompanied by a young boy. The monk explained that he would lead us in a brief "class," which ran for about ten minutes and consisted of a discussion of how to assume a "prayerful mindset" for the day. He then led us through a short series of yoga poses, basic stretches, and continued meditation. Our faculty group giggled as we awkwardly sought to balance the glaring contrast between our flexible, fit, and soft-spoken monk leader, and our own physical limitations. I felt among us an almost tense desire to be respectful, if not reverent, but an impulse to make jokes that would break that same tension seemed to leave the monk unimpressed.

Following the yoga session, another monk led us to a large lecture hall. The structure was built of compressed earth bocks, and an outdoor corridor wrapped around its full length, providing shade and enabling a sense of constant contact with the outside. Once inside the lecture hall, however, the atmosphere was dim and poorly lit. We took our seats.

The next two hours saw the lighter emotions of the early morning turn to a clearer frustration. For the full session, a speaker delivered a lecture that seemed completely disconnected from the theme of the visit. Entitled, "Overcoming Anger," the content focused on techniques of anger control and the ways to practice "the art of happiness." The claims made throughout the lecture were vague, and left unattached in any explicit way to a specific philosophy.

We were all aware that Govardhan is an ISKON undertaking, so in some ways we all expected some amount of orientation to its philosophy, especially insofar as it was related to the story of the site itself. Yet we sat in a lecture hall, left to gaze outward toward the environment and the larger grounds, feeling as if we might be experiencing some kind of nonspecific recruitment.

During a short tea break and a moment outside, we shared brief conversations and affirmed our puzzlement. Yet we'd only begun the weekend program, not to mention the day, so we returned to the classroom with the shared hope that the topics would soon shift to environmental architecture, or at least consciousness as it related to good design.

The next lecture was called "Stress Management," and as it went on the faculty grew, somewhat ironically, restless and impatient. I wrote in my notebook that we all seemed increasingly stressed out; we had yet to discuss anything overtly related to environmental architecture. Like the others, I felt distracted by a keen desire to go outside and explore the grounds. When the lunch break finally arrived, collective relief followed us quickly out of the hall and into the larger village.

In the afternoon we made our first guided visit to the larger eco-village, beginning, unsurprisingly, at the extensive cowshed. The head of the cowshed greeted us to explain that the core mission of Govardhan, and a pivotal aspect of its spiritual and ecological philosophy, is cow "protection." The cowshed structure was therefore a kind of dual center: it served both spiritual and sustainability goals. He continued to explain that, with the right number of cows to balance the human population in the ashram, human needs for food and energy could be met sustainably. While Govardhan had not yet achieved full self-sufficiency, our host told us that its ultimate objective was to provide for all of the village's needs, and to process all of its wastes. In this way, ultimately, the village sought to operate with a fully closed nutrient loop.

By this point in the day our full group was assembled, and we represented a wide range of faith traditions and identities. Among us were Muslims, Jains, Christians, and Parsis, as well as Hindus whose identification with Hinduism spanned a broad continuum. We were well aware of the profound symbolism of the cow in Hindu tradition, but also acutely mindful that an exclusive and often violent mobilization of orthodox ideas of cow protection was active, to tragic effect, in many parts of India.

Although we had not yet heard any direct references to ISKON in the morning seminars, touring the cowshed and walking through the eco-village reinforced the ideology of the devotees we were among. The same could be said of Auroville, to be sure, insofar as its members are also devotees of a specific spiritual philosophy and members of an ashram. Yet the dual fact that our program had, to that point, featured very little explicitly environmental content, and that the aesthetics and practices of the ashram created such a clear distinction between us as visitors and our hosts, the monks, gave this study tour a discernably different valence. Those present in Auroville tended to dress differently from one another, for instance, creating a general feeling of relative cosmopolitanism. By contrast, the Govardhan residents with whom we interacted were all men, and nearly all were monks dressed in white, saffron, or red robes. Even this simple uniformity marked us as unmistakable outsiders.

When I asked an RSIEA faculty member what she thought of the site at that point, her reply came slowly, as though she was seeking a way to be both respectful and appropriately critical. She offered that the most important thing to her was that students coming to Govardhan would be exposed to an alternative model of living, one that formed a social basis for ecologically sound commitments. "Even if the efforts are not perfect," she said, "the experience of being in a place that is organized so differently is extremely valuable. Most of the students have never been to a place like this," she said, and the exposure to it would attune them to what is possible in the world. Although it went unsaid between us, there was an overt feeling of orthodoxy and extremity in Govardhan that, for a variety of complex reasons, we had avoided confronting in Auroville. Here, the premise of an expansive and inclusive "Indian" identity, which linked to the consciousness-good design duality, seemed far less tenable. At the same time, faculty members repeatedly expressed to me their confidence in the students' capacity to engage the experience critically and selectively. In other words, faculty members felt that there was still a crucial, politically neutral message about sustainability embedded in the experience of Govardhan, and it was that environmental consciousness that made this study tour valuable.

Later, the same faculty member confided that she was not impressed with some of the underlying spiritual messages that came through as the workshop went on. "At ISKON, they start with shared values," she said, "and so in some ways meeting the challenge of running an eco-village is easier." Furthermore, the affinities between members of the ashram were not fundamentally anchored to environmental values, *per se*, in contrast, perhaps, with the experience at Auroville.

The second day involved extensive tours of the grounds and built structures. A session at the main assembly hall highlighted its interconnected environmental features: constructed of compressed earth blocks, illuminated largely by natural day lighting, cooled by low-energy cooling technologies and powered by solar, this was a building that seemed to fulfill a long list of desirable environmental attributes. We learned here that vermicomposting, on-site wastewater treatment, and the use of biogas all figured in the environmental supply-waste chains throughout the eco-village.

On the mid-morning tour across the village grounds, our monk guide began to connect Govardhan's built form to a narrative of "Indian" history and identity. "You see," he said, "the traditional lifestyle was logical and eco-friendly. We want to show people that this lifestyle was not for a lack of knowledge; instead, they had great knowledge that was lost when we became an industrial society." He stopped walking, and addressed himself to the full group:

> Stress is the number one killer of people today. Why? Because we live in a greed culture. We have lost connection with Mother Earth. Since the Industrial Revolution she has become a wife to be exploited. Everything we needed to know we knew in ancient times, before industrial society broke our relationship to the earth. For us, the earth is Bhumi Devi. Our goal is a harmonious life.

In some ways, this logic resonated with a similar set of points we'd encountered in Auroville, yet again, in this guise the group was both interested and uneasy. Though the logic of recovering values and technologies humanity had known through deep history was consistent, the inclusive potential to claim that history as one's heritage did not. Here, "Indian-ness" was a much more difficult category to disentangle from a reflexive awareness of the sociopolitical life of this same logic.

The group shuffled onward for another visit to the cowshed, and the monk stopped before entering to deliver an extended talk about the "basis of the Vedic lifestyle." "In traditional society, he began, people depended on two elements: cow and land." Cows and land formed a "perfect circle" for fulfilling the needs of human life. The manure was a source of fuel, and a rich form of soil nutrition—that is, a natural fertilizer. Unlike other forms of excrement, which the Vedas deplore as polluting, cow manure, he told us, is revered and "auspicious;" it is a soil nutrient that long predates chemical fertilizers. It is also the foundation of biogas, which powers much of the ashram, and an ancient disinfectant. He continued to describe the use of dung for plastering floors and walls, and its extensive use, along with cow urine, as an antibacterial agent in Auyervedic medicine. And of course, he said as if in an afterthought, cows provide milk.

In a switch from historical tradition to contemporary technology, our guide encouraged us to "see for ourselves," not only here but by consulting Govardhan's extensive website. Much later, curious to follow his advice, I found the ashram's extensive archive of articles, essays, and other materials. Among titles such as, "Land and Cow: the Green Miracle," and "Land and Cow: a Perfect Sustainable System," I read:

ONE ONLY NEEDS TO TURN THE CLOCK BACK BY FEW DECADES AND GLANCE OVER INTO THE LIFESTYLE OF TRADITIONAL INDIA.

Life in traditional India was purely centered around the culture of fulfilling one's needs and not one's unlimited wants. And two things played a vital role in setting up such a sustainable lifestyle—Land and Cow. Unlike the modern Industrial systems, Land and Cow form a perfectly sustainable system that can fulfill the needs of the human society. The output of a Cow barn is manure, which is a first class organic fertilizer and acts as an input for the agricultural land. And the output of the agricultural land, the grasses and fodder, act as an excellent Cow feed, thus supporting the Cow barn. Thus we see that these two systems are self-sufficient sustainable systems that can go on if properly taken care of. And the byproducts of these two systems namely milk and other dairy products, electricity through biogas, various cosmetic and medical products from Cow dung and urine, vegetables, fruits and grains are essential for mankind's survival.[30]

Once again, we entered Govardhan's well ordered, kept, and pleasant cowshed, and for over an hour, students and faculty wandered, almost adoringly, through the structure. Many posed for pictures with the cows, and the entire group seemed to linger to enjoy being among these impressive, and undeniably beautiful, animals. Our urban origins compelled some slightly awkward, and sometimes sentimentally charmed, giggles as students and faculty moved from shed to shed. Some gestured as if to lovingly pet the cows; others recited all of their "cute" attributes. Among the animals, the sociopolitical sting that accompanied the phrase, "cow protection" nearly vanished.

The everyday reality of tending, grooming, feeding, cleaning, and caring for the cows was not the visit's emphasis; the shed was instead a spectacle meant to animate the romantic story of cow-human harmony, and to provide an opportunity to revere and show gratitude to the creatures. Our monk's comments framed the animals as actual repositories, if not guardians, of an aspect of the ancient wisdom of good design.

We continued to the ashram kitchen, where monks offered explanations of the village biogas system and a chance to hand-churn milk. The program concluded at the site's extensive waste-treatment and plant nursery complex, called the Soil Biotechnology Plant. Here, yet another monk offered a slightly modified description of the perfection of "traditional" Vedic lifeways, hinting at their social complexities and exclusions: since in the past, human waste was collected by a particular social group (which he did not name), he told us, and since this work degraded and marginalized that group, the Govardhan community recognized the need to manage the human waste at the ashram and eco-village in a different way. The monk explained that relying on water as a basis for waste transport and decomposition systems was inefficient at this site, so the Govardhan model relies on soil microbes as the principle decomposition agent. The ashram's waste treatment system and its extensive plant nursery are in this way interlinked, and comprise a large complex that, like a composting toilet, creates plant fertilizers from human waste inputs.[31]

The comment at the Soil Biotechnology Plant offered the group a passing, but nevertheless overt, acknowledgment that the social systems that accompanied the version of Vedic lifeways we encountered at Govardhan had not been perfect across the long sweep from ancient history to the present. Yet the experience of the site remained qualitatively different, and in the end was considered far less "effective" than the study tour to Auroville for imparting RSIEA's pedagogical vision of good design. After the trip ended, faculty assured me that it was unlikely that future environmental architecture student groups would return.

Govardhan offered both contrast and consistency, then, to Auroville. Constant, clear referencing of Vedic "wisdom" and "tradition" made it impossible to fully separate the contemporary symbolics of Indian national and Maharashtra state politics from our experience of this place, and therefore almost impossible to glean an appropriately expansive and accessible notion of Indian-ness that could strengthen good design. It was far more difficult to encounter Govardhan as a place where the environmental objectives and values that its residents espoused could be regarded as somehow secular or secularizing, to say nothing of universal. The core message—to seek in precolonial history the remedies of all postcolonial modern problems—fell flat and unconvincing at Govardhan, even if that same logic had enjoyed traction at other moments in RSIEA training, and in other settings. Perhaps here, it was impossible to simply foreground environmental reflexivity, suspend sociopolitical reflexivity, and proceed to build a sociality with its basis in a shared idea of good design.

. . .

FIGURE 11. Students and faculty on the field study tour of Govardhan Eco-village listening to their guide describe the sustainability features of one of the site's main buildings. *Photo by the author.*

Ever since the classic work of scholars such as Wolfe (1982), Hobsbawm and Ranger (1983), Anderson (1983), and others, political ecologists have taken interest in the ways that ideas and practices conventionally accepted as "traditional" and "historical" are invented, and reinvented, for the time and context within which they are mobilized. Each historical element invoked and retold is aggregated for the present, fused to power relations and specific sociopolitical objectives that may reinforce, or may aim to contest, dominant patterns of power and authority. In matters related to environmental change, we often invoke scientific knowledge and environmental "awareness" to inform a list of *choices* for action, but it is through dominant and contested historical narratives that social actors assign meaning and stakes to the environmental actions they undertake. Nature is not only made, then, but also made meaningful, in part through these everyday retellings of the histories and identity attributes that matter.

Any version of a historical basis for a concept of Indian-ness that accompanies a concept of good design carries with it an implicit politics, made known through what is explicitly identified and narrated, but made equally important by what is left unsaid, who is excluded, and which issues are simply ignored. In the particular case of architectural form and nationalist narrative in India, this is well demonstrated, and has been convincingly argued, across a range of colonial and postcolonial cases.[32]

RSIEA's environmental architectural agenda, particularly when it came to building an environmental affinity group, required a historical basis for sustainability, a social capacity for broad-reaching inclusion, and a set of meanings that gave environmental architecture its urgency and purpose. The wide range of backgrounds and identities students brought with them prior to the program needed a singular, coherent ideology of belonging, not only to forge a strong collective, but to inspire that collective to move from theoretical commitment to actual praxis. RSIEA's narrative of Indian good design and its history could not afford to marginalize its practitioners, and thus required a nominally inclusive history to which all students could stake legitimate claim. The necessary silences, however, sat uncomfortably with a sense of complicity, precisely with the politics of exclusion and marginality.

Here, we might observe a contrast between the experience in Auroville and that shared in Govardhan. In Auroville, it was possible to embrace environmental reflexivity without having to undertake, at least in any explicit detail, an attendant sociopolitical critique. Here, the fallacy that a politically neutral and removed "environmental" stance was possible underpinned a sense of Indian-ness that could be accessed by every environmental architect, regardless of her background. Yet, as noted above, that same fallacy was undone in the disquiet of Govardhan. Its socio-politics were too overtly exclusive to enable the neutralizing environmental imaginary; the eco-political disconnect was never dismantled, only in brief moments suspended.

Neither study tour reached a more theoretically satisfying level of engagement with questions of history, identity, Indian-ness, or good design. That level would depend, as Sears has argued, on recognizing that " change and hybridity are ongoing processes, and that we should embrace the multi-layered nature of . . . tradition rather than sifting through it to find the truly pure."[33]

For Sears, one way to mitigate the silent exclusions characteristic of the study tours would be to simply seek and name the elements of hybridity and process. She writes that by embracing a multi-layered past:

scholars and architectural practitioners can provide strong resistance to the imposi-
tion of nationalist narratives onto material culture, which, in modern India, often
privilege the earliest moments—the Vedic or the Mauryan—over later periods. . . .
(By) recognizing that both Vedic India and Sultanese Janupur are part of the history
of contemporary India, we acknowledge that identity is composed of multiple layers
built up through human experience.[34]

Lest we leave this matter with the impression that the capacity, and indeed responsibility, to draw knowledge from notions of Indian history or Indian-ness is somehow unique to environmental architecture, or to the environmental anxieties of
the present, I conclude with just one illustration of the ways the issues explored in
this chapter are continuous with, rather than a break from, the past.

In the fall of 1987, the international publication *Architectural Review* published
a special issue on Indian architecture.[35] Its contributors attempted to survey Indian
architecture in terms of more "established," Western architectural thought that
predominated at that time. The introduction reads:

> Throughout the world, architects are attempting to evolve a contemporary archi-
> tecture that shows the respect for history and tradition that was abandoned during
> the pioneering frenzy of the Modern Movement and yet is capable of fulfilling the
> demands of late-twentieth century society and reflecting its aspirations. The search
> has, in the west, produced much grotesque and self-conscious architecture in which
> historic references are used superficially and ironically. But in India, where there is
> continuity between past and present, there is the promise of a more sophisticated
> and authentic synthesis between old and new and indications that a genuine archi-
> tectural future may be found by reference to the past.[36]

Thus the conventional international gaze has long attributed "continuity between
past and present" to India, and often ascribed to it an apolitical and deeply problematic orientalist posture. Yet one cannot ignore the resonance between the RSIEA
Opening Day assurances that, "It is Western nations that should be looking to *us* to
learn about sustainability; it is only India that can teach them inner growth."

The same volume of *Architectural Review,* includes an essay by William JR
Curtis, who calls Charles Correa, prevalent on the international stage at that time,
"pivotal" not only to "Indian Modern architecture," but also, potentially, the practitioner of an alternative form of praxis, one with the capacity to "synthesize old and
new" and in doing so "address what is pressing in the present." He writes:

> The best recent architecture in India may contain relevant hints for the develop-
> ing countries. It is becoming increasingly obvious that the uncritical adaptation of
> Western models is no real solution, as these are often inadequate to climate and cul-
> ture: the results tend to be alien and alienating. But the answer does not lie in the
> superficial imitation of local traditions either, as it fails to update what is substantial
> about the past, and does not address what is pressing in the present. The hope is to
> make a relevant synthesis of old and new, regional and universal. The best recent

FIGURE 12. RSIEA students exploring new construction in one of
BCIL's housing developments outside of Bangalore. *Photo by the author.*

Indian work is so challenging because it is open to the tests of the future as well as
the grandeur of the past"[37]

Curtis would go on to write a monograph about BV Doshi, in which he took great
interest in Doshi's work for its " . . . search for an architecture of authenticity based
on a philosophy of life."[38] This phrase captures, perhaps, one dimension of the
study tours as they charted an Indian basis for environmental architecture in the
present.

With the work of defining good design infused with experiential and classroom-
derived concepts, techniques, and modes of consciousness students would need,
and with a strong environmental affinity group made ever stronger as the two-year
program progressed, it became ever clearer that one critical element of the ecology
in practice puzzle was missing. The sociopolitical reflexivity that would allow environmental architecture students to operationalize the design approach that most
by now so strongly embraced was not, and would not be, part of the experience of
training. The chasm between aspiration and practice, then, remained potentially
vast and unbridgeable.

7

A Vocation in Waiting

Ecology in Practice

"Generally, I would say just going for (LEED or GRIHA) certification is not a great idea, but for Mumbai I would say actually, go for it, because something is better than nothing."

—AMRIT, RECENT RSIEA GRADUATE

"Metrics like GRIHA and LEED are not for the masters; they are for the followers."

—SHIRISH DESHPANDE, RSIEA FACULTY MEMBER

"I am very positive. Think of how Mumbai was ten years ago or fifteen years ago, and what we are, where we are today, it's good. Another fifteen years and there will be a lot of change. Maybe not a sustainable city, but we will be able to be environmental architects."

—SUHASINI, RECENT RSIEA GRADUATE

Upon completion of their final thesis, RSIEA students resumed their professional lives. Some returned to the smaller Indian towns from which they'd come, and some moved on to jobs in other large Indian cities. The vast majority of graduates, however, stayed in Mumbai and continued working in the firms or practices with which they'd remained during their studies. Newly conversant in RSIEA's version of good design and green expertise, they now faced the challenges of implementing their new approach in practice. This chapter addresses whether, when, and how graduates operationalized what they had learned. What did it take, I ask, to transform good design aspirations into actualized built forms?

This question moves beyond the observation that cities are repeatedly reimagined to point to the conditions that may enable certain forms of social action, and thus beget certain material forms. If this book began by addressing the lived social life of environmental architecture through its concepts, techniques, and moral ecological framework, it now arrives at the point of action. In this chapter,

I describe the experiences of a subset of RSIEA graduates to address how the very idea of the possible in good design was reconstituted, adapted, and actualized according to each architect's ultimate social structural position. How, I ask, did environmental architects produce and reproduce Mumbai's political economy in their efforts to promote environmental and social change? How and when did they make specific attempts to influence the city's built form, and to what effect? After all, both aspiration and operationalization are bundled in the concept of ecology in practice.

My aim is a better understanding of the contextualized meanings and relative power of RSIEA-style good design, as this was posed in relation to other active and powerful categories. In this chapter, important categories like "the state," "the builder," and "the architect" emerge in ways that contour the operational terrain of good design, and at times limit its capacity to "do" anything at all. Each marks a narrated concentration of power in Mumbai's urban development; each points to the layered institutions, political processes, and forms of knowledge that shaped or affected RSIEA graduates' specific suite of available choices, meaningful actions, and possible strategies.

The tension between aspiration and practice I trace in this chapter underscores a durable, and yet often pliable, balance between the compelling appeal of alternative, more environmentally vibrant urban imaginaries, and the deeply ambivalent relationship those who are embedded within them hold to a city's established political economic trajectory. I describe architects who acted from their specific social and political positions, each with dual stories of power and vulnerability, and each embedded in both the historical moment and the prevailing political economic realities of urban development.

Through many creative and conscious actions, RSIEA-trained environmental architects in part reproduced and in part refashioned that political economy. It is this process, animated by architects' narrations of choices, biophysical and social structures, and the categories of actors with whom they engaged we approach an understanding of RSIEA's environmental architecture as ecology in practice.

To follow, I examine how RSIEA-trained architects described the professional urban context within which they worked. I highlight descriptions of the strategic relationships they forged, and the moments when they named social, political, or material forces that seemed to limit or shape their practices. I draw from an initial data set of quantitative, descriptive statistics, written interview responses, and interview transcripts to assemble a collection of narratives. Rather than separating current students, new graduates, and their senior counterparts across set phases, the chapter is organized in terms of questions and themes that help to illustrate the challenge of transforming good design as aspiration to ecology in practice.

· · ·

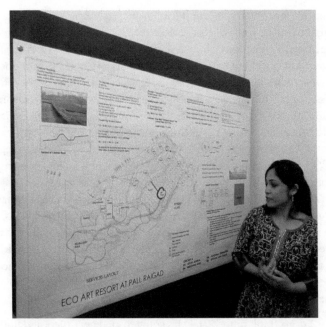

FIGURE 13. A graduating RSIEA student presents her team's final Design Studio proposal for an eco-resort at Pali. *Photo by the author.*

A basic assumption one might make when exploring the question of why a conventional architect might seek RSIEA training is that there is some kind of perceived scope for the future of environmental architecture in Mumbai. This, then, was one of the basic questions I posed to current and past students, usually to an enthusiastic and optimistic response. The following conveys one example; this offered by a student who was just completing her RSIEA training:

> There is excellent scope for green design, not only in India but also abroad. The issues of sustainability have touched the lives of poor and rich, young and old, everybody. So slowly but steadily everybody today is talking about being environmentally sensitive. Buildings coming up today want to get credited with LEED recognition, which is not enough but nevertheless represents a first step towards becoming environmentally positive. At the same time nature itself is ringing all the alarm bells. The 2005 Mumbai floods, and the global warming that has taken place have alerted everyone towards the wrath of nature.[1]

This particular assertion of certainty about an inevitable environmental shift, with an accounting of multiple signs that it was well underway, was offered in conversation with four RSIEA students facing immediate graduation. Of the three women and one man, all intended to maintain an architectural practice in Mumbai after graduation. Of those who speak in the following exchange, Palavi works in a firm of roughly fifty architects, which she described as busy with contracts for LEED and GRIHA certified designs, while Kalpana maintains her own private firm with one partner. Arash worked in a large, well known development firm prior to, and during his two years at RSIEA, but upon graduation had plans to start working for a smaller developer which he described as, "more sensitive toward the environment."

I spoke with this group immediately after the final event of the Design Studio course. The four had worked as a team in the course, and their collective proposal for the Pali eco-resort complex had just received a rather scathing critique by a small jury drawn from outside the Institute's faculty. Following a lengthy discussion of the points of that critique, I asked about their professional plans, now that training at RSIEA was largely behind them. Having established that all four intended to stay in Mumbai, with three remaining in their current firms, the conversation shifted to the development climate in Mumbai, and whether or not there is a scope for them to practice environmental architecture in a meaningful way there. As they discussed this question, a debate unfolded. The conversation turned to whether an environmental shift was eventual and inevitable in Mumbai, and the role of the architect in bringing such a shift about:

> *Palavi:* I don't think immediately it's going to change. Unless and until you get the client who wants green. Even if one developer is willing to do it, and if I start an initiative with them, . . . this is a chance. . . . The benefit to this builder is that he can market his work as "eco" and get a premium. He gets the money back that he put into it.

> *Kalpana:* Environmental awareness will take at least five or ten years to grow in India at the scale we are envisaging. But what I am targeting, personally, is there will be policies made by the government—they have to, and they *will* make them—policies that are nature-friendly, even for Bombay. In every industry, in everything. That will be the time when we, the people who have done a master's in environmental architecture, can step in and propose and work as environmentalists with those principles. Then the developers will choose us.

> *Palavi:* See, it's like studying computers. Ten years ago no one wanted to spend a lot of money to study computers. Now look. Nothing works without a computer, and those who studied computers are the economic leaders. Environmental architecture is going to make that impact. Everything goes through a phase, and this phase is going to lead to the next one.

> *Kalpana:* It *has* to. It *is* bound to. How much of land can you use up? How much can you cut down?

> *Arash:* By then it's too late, unfortunately.

AR: *But can you change this, as architects?*

Arash: We can change it, but we have to get into government. We have to be in policymaking.

Kalpana: I think every choice that you make, and I was telling Arash this a few months back, that we had been offered a project to do a housing colony in Panvel. They said we want to consume an FSI of 4 because that was a special regulation zone and they said we want this. And I said but why? That's Panvel! Why in that kind of landscape do you want to consume this kind of an FSI.

Palavi: I think you should have done it. At least you could have done it environmentally sensitive. If you say no, some other firm will do it and it will still be an FSI of 4, but with no environment concern.

Arash: They'll do it worse than you!

Kalpana: But for me it is about the impact that I make, not about the impact that I can reduce. It has to agree with my principles.

Palavi: The little interventions are what we can make.

Kalpana: But that's where we *cannot* make a difference. The moment we just pander to whatever is put in our way, then we are not doing it for the environment. We are just doing it for the fee that it's going to bring us. . . . So you have to make a call as to what it is that you will do and what you won't. The more people start making the right choices the better the impact will be.

Arash: That won't happen. Ever. Because people need money. They cannot survive and survival is the fees. Without that, nobody will survive. So? Why would you want to see the same site built in a pathetic way? You build it your way, you know, and at least a less destructive way. Just because I won't design it doesn't mean the work isn't going to be done. Someone is going to do it, and worse than you.

Kalpana: But you are doing something about it. I think every builder understands that any architect who is giving up that kind of a project and that kind of fee and that kind of money—because the fee is not small—the moment you say I will not do it, you are impacting the person's *mind*. He will stop and think that oh my god, that has to be something really strong that this person has said no to doing a project like this. It does not come by every day. So when you do that, that person will also. It will not be immediate but I think over time he will, it will impact him in some positive manner.[2]

This conversation underscores many of the issues that would recur as RSIEA graduates traced their experience once they left the Institute. On the question of the power of the architect, per se, to catalyze an ecological shift, the answer hinged on the relative power of incremental design interventions. In Arash and Palavi's logic, even "less destructive," albeit far from perfect, environmental designs were better than those that occurred in a purely conventional scenario. This logic took all incremental interventions as potentially valuable, echoing the RSIEA student quoted at the chapter's outset who acknowledges the imperfection of LEED and GRIHA certifications, but interpreted their growing popularity as "a first step toward becoming environmentally positive." For Kalpana, by contrast, meaningful and effective action requires environmental architects to refuse projects that they regard as fundamentally destructive and environmentally harmful. This was the counterpoint to the logic of incremental change, but to stake this position also depended on one's political economic capacity to do so.

In the balance of this tension is the willing forfeiture of economic gain, a calculus familiar in any debate over the appropriate politics of social transformation. In the students' exchange, the "disengage until the proper terms are met" position offered a sharp rebuttal: in India, those who will accept the project exclusively on the basis of economic necessity are many. Policy and market structures, in this logic, are the ultimate purveyors of the kind of change that will create more demand for good design, and therefore enable the form of ecology in practice they sought to operationalize. Less clear, of course, was when and how environmental or political conditions might force state agents to reform laws and policies, and then to enforce them.

Despite the tension, the exchange above also illustrates a shared confidence that eventually, as one declares, "there will be policies made by the government— they have to, and they *will* make them—policies that are nature friendly, even for Bombay. In every industry, in everything." In an important way, the messiness of the *how* questions are far less important in this exchange than the resilient, shared confidence in the inevitability of change. One may be left to hypothesize, perhaps, that the architects regarded the breaking points of ecosystems themselves as the ultimate catalysts for a clearer path to practicing environmental architecture. Human power relations, when cast in the context of global environmental crises like climate change, natural disasters, pollution, or habitat destruction, seemed inevitably susceptible to reworking—particularly if the agent of that reworking was the very environment itself.

In another conversation among another group of four graduating students who had just experienced their *Design Studio* final project critique, the idea of an inevitable environmental shift repeated. Rather than casting the environment as a likely agent that would force the change, however, the participants in this conversation narrated a more conventional, if still dramatic, political economic shift. The shift was well underway, my interlocutors assured me, and its indications could be mapped globally.

This was, again, a group of three women and one man; here all planned to practice architecture in mid-level Mumbai firms following graduation. A field notes excerpt from the conversation reads,

No one said they had joined (RSIEA) with the intention of putting environmental architecture into practice right away, but they described seeing change coming quickly, across the globe. Kabir spoke passionately, and at length, about an inevitable global revolution, saying it had already begun. He cited the Arab Spring uprisings, the Occupy Wall Street demonstrations, and the rise of groups like Anonymous as evidence that, in his words, "this kind of global capitalism is going to end in our lifetime." Naturally, he argued, what follows it will be more sensitive to the environment; it will support healthier cities and green design because it will value what environmental architects do. Not everyone was as confident that a total, global political economic revolution was imminent, but all agreed that significant global political and economic change is coming, and quickly. . . . To this confidence that global capitalism in its current guise is time limited, Aahna added an ominous assertion. She said that given how "chaotic" urban development conditions in Mumbai have become, they are bound to get much worse before they get better. Again, I asked why. Her reply was that "greed and profit" fueled all current urban development decisions in the city, and "you cannot take that down overnight." Aahna said she expects that Mumbai is going to implode before a revolutionary change takes place, or, she then reconsidered, perhaps they will happen at the same time. When I asked what she meant by, "implode," she talked about "suffering and chaos." Then a long silence fell over the group, until Kabir said, "Watch for the next Occupy movement. It will be somewhere in Asia and it will be even bigger. This is what I expect. They will keep happening and then there will be a huge shift."[3]

In light of this and the previous exchange, we might understand the training and practice of good design not only as a realm of ecology in practice, but also as a kind of anticipatory political and professional refuge. Training in good design was a way of ensuring that one had the architectural capacity to make "order" out of an inevitable and approaching urban environmental chaos—a positioning of the professional and personal self against broader forces regarded as both time-limited and self-destructive.

I was struck at first by the overt indictment of capitalism in this conversation, if only because it was so rarely discussed in any overt way during RSIEA training. The culprit in this logic had a shorthand name, capitalism, and the system that organized its social and environmental effects, and which also organized the patterns of urban development that kept Mumbai on a specific, environmentally undesirable trajectory, was withering in real time. The rather calm and casual agreement in support of an anticipated revolutionary change surprised me. Yet as my conversations extended beyond graduating students and engaged those who were now back in architectural practice, I found that idea of a vocation in waiting—not to mention on the correct side of an inevitable revolutionary change—consistently reinforced. The revolution may be socially- or environmentally driven, but it was surely coming.

Understanding this depends in part on the experiences and responses of differently positioned, RSIEA-trained environmental architects. In surveys returned by current students, the overwhelming reply to the question, "What is the greatest obstacle to environmental design in Mumbai?" was simply, "lack of awareness." This reply from a current student exemplified a logic that connects growing environmental awareness with an automatic social response: increasing demand for environmental architecture:

> It is heartening to see a lot of awareness campaigns and citizens' initiatives towards saving the environment and one hopes that these will bring about some amount of change, however small. Lack of awareness on the part of both clients and architects is the biggest challenge and obstacle. Bringing about awareness about the subject can go a long way in creating a demand-supply system for environmental architecture so this knowledge will definitely be helpful in propagating an urban lifestyle of low consumption.[4]

Among practicing RSIEA graduates, however, more complex descriptions of the political economy of development, its contours, and the nature of the changes that would have to accompany a shift to environmental architecture, emerged. Here, certain narrative categories organized depictions of the urban development present in Mumbai. These included social norms and processes, such as bribery and corruption, as well as specific actor groups, such as builders, government officials, and architects. These were all invoked as aspects of the lived "market," or what Kabir would perhaps call more overtly Mumbai's specific experience of global capitalism.

If more enabling structural changes were just on Mumbai's horizon, what, more precisely, was the timeframe architects described? How, and how soon, would essential urban development structures enable environmental architecture in Mumbai, and, yet again, what would catalyze it? Let us consider first comments offered by an architect whom I will call Siddharth, an architect who was completing the RSIEA program when we first met in 2008.[5] As seatmates on the long bus journey between Chennai and Auroville, we forged a continued interest in one another's work that long outlived the dusty, sundrenched heat of Tamil Nadu. It was there that Siddharth had first described to me his ideas for a set of bungalows that his boss was working on in Africa. On the bumpy bus ride, he sketched design elements on a scrap of paper, carefully itemizing how its many aspects harmonized with good design principles.

Since our very first interaction was framed by a discussion of this project, I had thenceforth assumed that Siddharth joined RSIEA because he had a deep commitment to environmental architecture. But years later, in the context of an interview in 2012, he assured me that this had actually never been the case. Instead, his interest in the RSIEA program was simply driven by a love of design; environmental architecture was a way to sharpen his sense of connection between built forms and landscape. Siddharth described his experience at RSIEA as a way to expand his concept of design, not a platform for forging eventual (or inevitable) environmental change. I learned in those later interviews, in fact, that he did not actually "bother" to complete a final RSIEA thesis when he finished years earlier because, in his words, "the knowledge was more important than the degree."

After leaving RSIEA, Siddharth shifted from the small environmental design firm with which he'd worked in 2008 to a much larger firm of seventy architects; his current firm undertakes projects all over India. Few of these, he told me, had environmental dimensions; when I asked why, Siddharth said flatly that the builders were not asking for them. Seated together in the open-air institute cafeteria in 2012, I asked Siddharth his working impression of the scope for environmental design in Mumbai. His quick, sharp reply caught me off guard, in part because I'd been focused in the days prior on interviews with current students, so many of whom expected at least some scope for environmental design practice upon graduation. He said:

> There is no scope for environmental architecture in Mumbai. For any project, the commercial aspect—the profit—is much more important than environmental impact or the design aspect. Because finally (builders and investors) want to earn . . . it's more about money than real architecture or environmental values. I don't think it's possible to change this until everything changes. Right now the politics and economics are completely against environmental architecture here. Only a whole new economy will create a scope to practice real environmental architecture and design.[6]

What was missing from the conversation with Siddharth was the confident assurance that such a "whole new economy" was somehow imminent. So long as builders and developers were profiting, he could find neither scope for good design nor promise of political transformation.

Conversations with other practicing graduates offered a similar view. Aditya, a 2009 graduate whom I'd first met during the study tour to Chennai and Auroville in 2008, told me when we sat down for a 2012 interview that he had initially hesitated to accept my request for an interview. He was eager to say hello after several years, he explained, but he was concerned, in his words, "because basically my experience is anti- your thesis." I asked him to elaborate on what he meant, and he replied that he understood me to be studying actual environmental architecture in Mumbai. Yet since graduating, he had been doing almost exclusively "anti-environmental architecture," not by choice, but out of the necessity of his work. "My work," he repeated, "is anti- your thesis."

Aditya works for a large, well known firm of over seventy architects; the practice is focused on mid- and high priced luxury residential developments in Mumbai. Several of Aditya's projects have high visibility and recognition in South

Mumbai, the most exclusive part of the city. During one of our 2012 conversations in a teahouse not far from the Institute, he described his dire view of the scope for environmental architecture. He based his view on his work experience:

> No, for residential buildings, I don't see any scope in Mumbai to practice environ-mental architecture. . . . I don't think it's really possible in Mumbai. Not only are the builders looking for their profits, but once it is built and lived in, in residential buildings everything is about personal consumption. If I am paying for my flat, then I think I can use AC for twenty-four hours a day. I can consume, and consume con-tinuously. Just that example, AC is normally on all the time if you're in a high rise building. In fact these luxury high rises have central AC. These are 3BHK and 4BHK. So really, where is the green design here? Not in the building itself, and then not in how people use it once they live there. There is no energy efficiency. No water ef-ficiency. Nothing.[7]

When I asked whether in the design and construction phase there was scope, at the very least, to use alternative materials or make other simple interventions (according to the "incremental change" logic discussed above), he replied:

> Like for meeting green building criteria? I'm not seeing this used very often in resi-dential developments. Maybe on paper or in their advertising, but no, the environ-mental architecture we studied in RSIEA, this is completely not possible in residen-tial development in Mumbai. In reality, by being a residential architect in Mumbai I am working in an anti-environmental practice. It's not what I was hoping for.[8]

RSIEA boasts a number of prominent graduates in Mumbai; among these is a well-known builder whom I will call "Contractor." Contractor employs huge teams of architects, and among them is someone I shall call "Darius." Darius met with me several times to discuss his experience as an architect in a large builder's firm in Mumbai. At his invitation, we spoke over tea on the open-air deck at the very exclusive Bombay Gymnasium. This private club in the heart of South Mumbai was originally established in 1875, when its membership was limited to the British. In the present, it remains exclusive, but its membership boasts a par-tial who's who of Mumbai's elite. "Bombay Gym," as it is often called by its mem-bers, is remarkable in many respects, but prominent among them is the relatively large plot of open, green space it occupies in the heart of South Mumbai. Shaded with dense mature trees, the plot's perimeter marks a clear boundary between the relative environmental calm of the club space and the dense, bustling city just beyond its borders.

From the comfort and relative calm of this place, Darius and I discussed his work experience. When I asked how it shaped his view of the scope for practicing environmental architecture in any sector in Mumbai, he emphasized the historical moment for urban development in India—not in the sense of an imminent new development plan, but rather in the context of the broader global economy. The relative economic boom in India, he said, would trace a predictable continuum in which a consumption-hungry urban middle class drives a specific phase of capitalist market growth. Referring backward to the neoliberal economic turn in India, he made the inevitable comparison to a prototype consumer society, and in doing so explained his sense that the current political economic order will not only continue, but that the state and market balance in contemporary India is in part to blame for what he viewed as a destructive culture of consumerism in Mumbai:

> America became a consumer society a long time ago, and India is going through that right now. We're doing it right and wrong. We're picking up a lot of the negative sides of consumerism much faster than we should. I think the beauty of India—at least with Nehru was that . . . (historically, the market sector in) India . . . remained closed. We industrialized and didn't allow any sort of foreign products in until I think 1990 or so. . . . I think economic globalization is good and fantastic and the way forward. However I think India needs to keep it in check. It needs to slowly introduce it, which is not the way things are happening now.[9]

He continued, connecting his reference to an emergent, bustling consumer market to the "way things work" in the urban development sector:

> These days, it's all a numbers game. The money is all that matters. At the end of the day if the client can save one lakh he will save one lakh. On one hand, one lakh means nothing to him in the context of these mega-projects he's doing for millions of lakhs, but then on the other hand it's a mindset. If he can save it he will save it. And if that means taking an illegal route, eight times out of ten, in my opinion, people will be fine with it. Unfortunately this is India. This is the world we're living in. It's sad.[10]

When I asked him to say more about what he meant by the comment that, "this is India," Darius echoed views voiced earlier in the book, in which the state and irresponsibly built form development are coproduced. His discussion of Mumbai, almost instantly pointed to the state of Maharashtra:

> The entire state is run by the Shiv Sena. . . . Take a simple thing like roads. Ninety-six percent of the BMC is Shiv Sena. The election just happened, as you know, and they won in a landslide. So for instance you have a simple road that gets built and does not last even for one monsoon. It falls apart. Then the building contractors are not held accountable . . . because the problem is that the building contractors are also Shiv Sena. So they're not going to pull them up. The political groups protect their own, and the sense that we all belong to Bombay, or to Maharashtra . . . is totally missing.[11]

A 2007 RSIEA graduate whom I will call Amrit worked with "a typical commercial architect firm" until 2010 and then established a private practice that allows him to work with individual clients on small projects "according to environmental principles," explained to me that while he is enjoying using some of his green design techniques in this new practice, he regarded the scope for any kind of environmental architecture in Mumbai to be limited to small scale projects within a very small arena of affluence.[12] It turned out that all of his current projects were in that scope, each for a client who wanted "eco-friendly" bungalows for weekend escapes from Bombay. These projects were in Ali Baug, a common destination for Mumbai's elite to establish second homes. When I asked him to describe his experiential impression of the scope for environmental architecture in an interview at RSIEA,[13] Amrit explained,

> Basically Mumbai has a lack of space, so nobody's coming to an architect to design new construction. You simply can't apply all your thoughts or work the way you *want* to practice. . . . It's completely builder governed, this industry. I would say in Mumbai, it's probably because they want to sell the property in a very limited span of time. There are also redevelopment projects and they are booming. But there you have that ratio of sellable areas and so within that constraint it's very difficult to achieve any of the environmental design aspects you've learned in the (RSIEA curriculum).

Where Amrit did find scope to practice RSIEA-style good design was in making precisely the kinds of interventions that were so heavily critiqued as inadequate back at the Institute. In the context of Mumbai, they were "better than nothing:"

> Generally, I would say just going for (LEED or GRIHA) certification is not a great idea, but for Mumbai I would say actually, go for it, because something is better than nothing. At least builders are beginning to recognize the value (of certification) for marketing purposes . . . so to a certain extent it is helping to save our environment. It's not ideal. It can be completely impossible in some aspects to practice green architecture here.[14]

A survey quotation from a recent graduate underscores Amrit's point about finite space in Mumbai and the work of the environmental architect, offering yet another complicating point:

> A lot of design is governed by development control regulations and though it is necessary, in many ways it does not give any design flexibility to the architects. Clients are always demanding that extra inch more. Plots in Mumbai are very small, and where every square inch has great value, any architect's focus lays much on consuming the entire FSI and so many times the focus on sustainable issues is left out.[15]

Ideas of spatial, political, and economic restrictions on architects' work often connected to descriptions of a cultural sensibility that animates the urban development process in Mumbai. Its primary characteristics might be mapped back to a portrait of the city's contemporary political ecology of urban development, but many architects emphasized a world of associations and meaning-making which they also attributed to image and marketing.

Particularly in the residential development sector, several interlocutors emphasized that despite what seems to be a proliferation of LEED and GRIHA certified projects and buildings, "actual" environmental architecture was virtually nonexistent. Both Siddharth and Darius repeatedly used the term "gimmick" to describe what, on the surface, can appear to be a proliferation of green-certified developments in Mumbai. Siddharth described one of his current team projects by telling me of a developer who approached his firm with a request for an LEED gold building design. After working on the project for a few months, Siddharth was convinced that,

> Right now it's a trend—a marketing gimmick. This builder doesn't actually want a green thing happening, in the way we think about good design. He doesn't care about the ecology at all. It's just a marketing thing. He's like most developers, I think. They know if they say it's LEED, it will affect their final sale.[16]

Darius substantiated this point to some extent, when he described the "green philosophy" of the builder for whom he works. Rather than signaling principles of environmental architecture or urban ecology, Darius explained that "green" quite literally often means adding green-colored things to building plans. He explained,

> Look, the way we think about green is in a pretty naive sense. We put greenery on our buildings. So for instance we build terraces. We (plant) trees. We build beautiful, but artificial, landscaped podiums. For instance in my township, my 300 acre township in Pune, in my plans I showed a bird sanctuary. Because it's a gimmick; it's what the builder can sell to his public, saying, "we have an urban forest." They are the cool words we use; I don't think they're true. When we say bird sanctuary, we just mean that birds will be there, or birds will come there. We're not actually developing a *bird sanctuary*.... It's a term we use to draw people in. It's a gimmick. So that's the obvious way we show green.[17]

Even as we spoke, many parts of Mumbai were plastered with billboards promoting new residential developments that promised "green bliss," a "green lifestyle," and "green luxury." The impression these boards and their attendant advertising campaigns created, if only through their ubiquity and visibility, was of a luxury residential sector that was literally transforming. With these new developments, they seemed to convey, the city could finally provide the discerning Mumbai buyer with the tranquility, efficiency, and general moral ecology of environmental architecture that he had been so desperately seeking. At very least, even I had assumed that these buildings were securing open spaces (albeit likely private and highly exclusive), vegetation, and, importantly, the infrastructure for an energy and water efficient domestic and service sphere.

Nearly all of the architects with whom I spoke convinced me otherwise, but none with more chiding than Aditya. He was unsurprised to learn that I had gleaned this impression, but he was quick to reform it:

AR: You are saying there is absolutely no authentic interest among any builders to erect green residential structures, but when I drive through Bombay, it seems that all I see are billboards advertising the new green luxury buildings. What's going on?

Aditya: Well, where is that building? (laughs) *What* is that building? Of course they are making all kinds of green claims, but there is nothing green about the actual buildings except the pictures on the signboards. I can tell you with my experience of trying to make even little changes that are green, in the residential sector there is nothing truly green. Yes, in the commercial sector you see a lot of push for certification, but even there it is about green cents (spells out c-e-n-t-s). There is no motivation to think in terms of environmental architecture, and it's even harder to imagine the residential buyer wants to use less water or energy or AC. Okay, in the residence they want some green lawn or something. But they will put a gate around it; they will use chemicals to keep it green . . . they will be anti-environment. This is why I am saying that my work is actually anti- your thesis.[18]

A collection of key actor categories consistently organized architects' narrations of the political economy of urban development in Mumbai. Here, the popular image of all urban development professions, the relationship between government and private sectors, and the ultimate room for choice and action which architects ascribed to themselves made the optimism and imperatives of responsible action that so characterized RSIEA's good design ethos seem almost ripe for caricature. These descriptions echo, nuance, and in many ways return us to many of the basic points to which Laxmi Dashmukh alluded much earlier in the book. Each came, however, with the personal narratives that brought these categories to life as facilitators of, or obstacles to, doing environmental architecture—structural obstacles to ecology in practice. I turn now to some of the descriptions of choices architects faced, and discussions of how those choices resonated with their ideological imaginaries. These connected to the deeply personal, moral, and even familial logics through which RSIEA environmental architects described professional compromises between RSIEA training and ecology in practice.

Perhaps the most prevalent figure in nearly every interview was that of the builder. Roundly despised, and often pointed to as the source of urban disorder in all its forms, the popular image of the builder in Mumbai is unquestionably negative. At the same time, many of India's wealthiest and most powerful figures are themselves builders. Even as they may be regularly critiqued in the press, in activism, and in many aspects of everyday Mumbai life, the names of the city's most prevalent builders are as known as any famous media or political figure, often with a mix of reverence and disgust.

Darius works for one of these prominent builders, and so it was in conversation with him that I was particularly interested in narrations of builders. One of our conversations took place on an evening when the day's newspapers were saturated with reports of a threatened strike by the Builders' Union. Our discussion thus turned to his views on the strike and its potential, but this quickly turned to the broader role and image of large construction and development firms in Mumbai.

When I asked how he felt working for a firm with such a renowned, but also somewhat notorious, figure at its head, Darius shrugged. "Well, he's a builder, and . . . nine times out of ten, I think, if you ask people, "What do you think of a builder?" they'll say, "they're corrupt." But in his view this image was unfair; whilst the processes by which construction and development proceed in Mumbai are anything but transparent, builders are nevertheless facilitating the provision of material development, he explained. "At the end of the day," he said, "(builders) are doing a service and without them the nation would be pretty lost. (Builders) are at the forefront of everything, in the sense that without them the city wouldn't work."[19]

Expressing some confusion about what, precisely, a builder union was, not to mention its relative power, Darius replied:

> Everyone talks about a builder lobby, but it's really a big bad ghost because they don't stand up for each other; there is no unity among builders. (I'm talking about the main firms, like) Lodha, Hiranandani, Raheja—these are the old names, and then you have newer ones like Peninsula. They were the ones who did Phoenix Mills, so they revolutionized the entire building industry. They managed to convert one of the old mills into a huge commercial district and this had never been done before. . . . It was amazing how they did it. The mill lands had been under litigation for thirty-odd years or forty-odd years and somehow they managed to twist the system, get the workers compensated—not well, but they sort of got the problem to go away—and then they built Phoenix Mills.

Hoping to probe the idea that the well-known, and comprehensively studied, case of Phoenix Mills could possibly have undergone such an almost magical transformation, I asked "do you mean to say that *no one* knows how they did it?" He continued:

> Look, there are a lot of loopholes. Indian law is written with vague intentions, so it's completely up to interpretation. And I think it's also partially luck. So for instance, if you can get the right officer to interpret (a law) in a certain way, it's all well and good and your project gets through. If you don't get that—and an honest officer is a joke; there's no such thing as an honest officer—you're basically working around the system. And the system is set up for you to work round it. If you try to do something through the legal channels, it will never get done, and the system is set up that way. You accept it, you move on. If you don't accept it, you don't survive.

Darius' response reinforced points that were echoed repeatedly in conversations with environmental architects: the laws and regulations are never the actual medium for effective action in urban development, and "the system" is actually designed in a way that invites "interpretation" and requires the capacity to shape that interpretation. To learn to see that system not only as it was, but to "accept it" and move in accordance with its choreography, was not just about practicing environmental architecture; it was about survival itself.

"What use is a Builder Union, then, if the process depends so much on subjective interactions with officials?" I continued.

> Well, for example, they (the Builder Union) have been trying to get rid of the new BMC commissioner. But it hasn't worked. Everyone claims that the Builder Union is the strongest lobby in India and they push and change things whenever they want to. But see, it's not the case. I think the Indian government has the ability to push back and keep them in check. The beauty of (his employer and head of the firm) is that he knows whose ears to whisper into; I think we get a three month warning before something is going to change, and we plan accordingly. So normally none of our projects get stopped, but even *our* projects over the last year have been stopped. This is really unusual. Under this BMC commissioner, there has been no leeway. . . . The stoppage is for all the violations that just a few years ago were pretty standard. What used to happen was if you had done something illegal, and you were caught, you would be fined. But then you would be allowed to continue. This has stopped.

What has changed, I asked, to allow for such a dramatic shift in enforcement norms? Why was the Builder Union having such a difficult time realizing their goal of "getting rid of" this most recent municipal commissioner?

> Yes, every municipal commissioner who has stood up to the builders in the past has been thrown out. But this time, I think he just has (the central government in) Delhi behind him. They claimed he wasn't going to last a week, because in the past they haven't. In the past they have been transferred out. But now I think the Congress government has had so many scams, between the Commonwealth Games and the Adarsh scam in Cuffe Parade, I mean they recently arrested a few people for that—and high up people. So he seems to be honest; they claim he's honest. And in India you can get that: at the head, yes, the honest are not corrupt, but he cannot control his fifty other minions that are below him. Every one of those people is corrupt.

The scuffle with the Builder Union, then, was symptomatic of a larger shift in assertions of power at the level of state and municipal government, one that might ultimately rework a calculus of state-builder power, but that would still, presumably, limit the work of the environmental architect.

But Darius' very deliberate assertions about "corruption," even as he described an entire system that works in a known, albeit not formally scripted way, compelled me to ask further about his experience of the way urban development actually happens in Mumbai. He took a deep breath, paused, and began a description that immediately introduced another important actor in many conversations with environmental architects: the municipal architect. He explained:

> To start a project in India, step one is you have to go to a labor commissioner and get his permission. That process, between fees and bribes, costs you 25–30 lakhs, depending on your project. Then you have a whole host of other officers you have to go through—the feeding order—and so that's where your municipal architect gets

involved. At (Darius' firm) we do not get involved in that. We are aware of what is happening, of course, but we don't deal with that. We don't deal with payoffs or making the bribes. So that's the job of the municipal architect.

Here, Darius made another common assertion: although this process exists, he explained, they were kept at a distance from his firm. Any necessary payoffs and shady dealings, he claimed, were in some way outsourced to municipal architects. He continued:

> You have a bunch of municipal architects; generally they are all ex-government officials. So they've retired from government and set up these businesses. They have left the service, but they have all their contacts and so keep getting their share of the pie. So it's sort of this circle of, after you retire, after you leave that department, you can retire or you can start your own practice and depending on how high you were up there, you have the connections and you have the ability to get things passed.

Darius had set the category, and its place in the processes of urban development, that allowed the architect to be simultaneously a part, and to stand apart from, the broader system he described. When I asked him to explain further, he described how a municipal architect functions in his own firm:

> The way the system works is that any building that is built by (Contractor) in Bombay, on paper his associates are not the architect. On paper you have the municipal architect. So if we get into trouble it's the municipal architect who gets into trouble, not (Contractor). This is standard.

Consistent with my conversations with environmental architects in many kinds of firms, Darius explained what is "standard" according to the established "system." Its dimensions were narrated as structures that confined action and limited available choices for how an environmental architect (or any other architect) can operate. In this sense, the municipal architect was rendered a rather neutral figure—a facilitator of other processes rather than, for instance, a derided subset of an otherwise respected profession:

> You have to have a municipal architect because drawings have to be done in a certain way. They know the rules; they know the loopholes. . . . So for instance a client hires this municipal architect. The municipal architect knows the ins and outs of the government, and knows how the system works. The municipal architect will tell him, for your project, roughly, it will cost you this much. Officially it's this much, but unofficially you will have to pay him X. The client then gives him a ballpark, saying, I'm alright going up to this level, but anything more is not feasible for me. So the municipal architect will go off and deal with the officer. Well, he'll never deal with the officer directly; he'll deal with the secretary or the appointed person. And generally the deals really do happen in these hotel rooms where they go, drop the money, and his fellow calls up and says you've got it . . . I've never experienced that, thank god, but I know how it works in that sense and I've heard of how things happen.

I asked Darius how he could possibly avoid experiencing this, given how closely he works with the firm and its projects. "Well for me," he said, "it's all still pretty shocking. And it's sad. Being my father's son, I know I cannot do anything illegal. I know it; I mean, I will get caught. So I have figured out how to avoid it and still do interesting work. But it's not easy. The system is set up this way."

In every interview I conducted among practicing environmental architects, the default position in discussing how Mumbai's urban development system works was to outline the technically legal activities in which the interviewee took part, and to describe an arena in which a particular subset of architects undertook any necessary illegal or "corrupt" tasks. Designating two separate spheres not only facilitated claims to practicing without legal compromise, it also reinforced the notion that a powerful structural system determined architects' capacity for action, rather than the other way around.

This was nearly always supplemented by some overt expression of disappointment, and occasionally disillusionment. In Darius' case, the municipal architects' material gains were a kind of index of injustice:

> What's really sad is that a municipal architect earns more money per square foot than a regular architect. And his fee is not including the bribes that have to be paid. His fee is just telling the client the roadmap . . . as in, this is who you need to give money to, and when. And he does the municipal drawings, which are totally different from what an architectural drawing would be. They are essentially two different projects. They submit one project (for permissions), but a completely other project is being designed. They are literally like two separate buildings, two different projects. At times it's mind blowing. I've seen some municipal drawings and I'm shocked at how our plans are transformed into this other set of documents. And these people are making so much money.

The challenge of maintaining a personal sense—or even a consistent and coherent narrative—of professional integrity in the midst of such uneven and opaque norms was a recurrent theme among nearly every RSIEA graduate with whom I spoke. In responses similar to Darius's, interlocutors spoke of drawing certain boundaries around the work they were willing to do. Some also spoke of a conscience, or in Darius's case, a familial sense of honor, that prevented them from engaging in the practices they labeled as corrupt. And yet even as each of them described their relationship to "the system" and "the process," it was clear that they, too, had probably had to compromise those boundaries at times.

Two graduates whom I will call Suhasini and Prisha are, like Siddharth, 2009 graduates of RSIEA. I have known both since the study trip to Chennai and Auroville during which I first became acquainted with Siddharth and Aditya. Both women joined the RSIEA out of a deep commitment to what they described to me as their "environmental values."

Thane-born Suhasini works for an urban planning and advocacy firm that is focused on "sustainable transportation systems" in Mumbai. She also lectures occasionally at RSIEA, and so in a short period has moved from graduate student to adjunct professor. Her speech becomes most passionate, however, when she speaks of bringing about positive environmental change in Indian cities.

Prisha, by contrast, works in a small design firm of four architects, with a changing roster of draftspeople and interior designers. Most of their work is in architecture, and Prisha described her experience in the firm with great enthusiasm. This is not a nominally environmental firm, she said, but "if a client comes to us wanting something green, we encourage it." Any lack of environmentally-focused projects was not, she said, due to the firm's collective willingness, but to the problem of demand.

In a comment that instantly challenged the overwhelming impression among current RSIEA students of a widening scope for environmental architecture in Mumbai, Prisha told me simply, "I have never had a single client come to me and ask for something green." She continued, "I think certification is still growing. We (her firm) are also relatively small, so if it's a big project and they want to make it green, then often that kind of client won't be pitching to us; they would take it to a bigger (firm)."

Although professionally positioned in different sectors, both Suhasini and Prisha emphasized the need to maintain boundaries around their work that allowed them to avoid engaging in illegal or "corrupt" practices and remain "neutral." When I asked how one could possibly manage this in the context of "the system," Prisha not only described a category of practice similar to the municipal architect, the "local architect," but she further differentiated her own practice from the wider "system" by emphasizing the aspect of the overall process in which her firm was active:

> In my firm we are lucky, I think, because we can just concentrate on design. When it comes to passing permissions, the client has his own local architect to do that. If we get into that, because we are a small team, it's like diverting your specialization in design. So this is so much easier. We all agree, let's stick to it (design); we specifically don't get into it (permissions).

But working on just one part of the larger process also created limitations. She explained, "this means that we are also restricted. But we have tried to create a line so we don't cross into politics because, I guess, it's not me, but my seniors must have indulged into it and (had bad experiences) so they have drawn a line as well, saying, no, we won't do anything illegal."

As Prisha talked through the details of this "line," Suhasini nodded her head constantly. Without prompting, she substantiated Prisha's expressed need to avoid illegalities at all costs by invoking the experience of her father, also an architect:

It's so bad otherwise. My father was working on a very prestigious project that was organized by WHO (the World Health Organization); they were designing disaster assistance schools. So I was working with him before taking a break to start the master's program. . . . For the first two years it was going well, and we were working, but my father was not getting paid. And the thing is, we were designing schools, and it was through disaster assistance, so the buildings had to be from good quality material. You know, no corruption and all. So he said we'll find a contractor, and we'll have a proper tendering process. But (the client) said no, they wanted their own local contractor who would also eat money like them. So then they stopped paying him. Even to this day, he has not been paid. So finally my father got so frustrated—after three years (they're) not paying (him) and this is too much. So he filed a case against them and it's (pending). When I see my father going through all this, I think, you know, why be in Bombay? I feel so discouraged when I see this.

Suhasini's story was in no way unique, and comparatively rather benign. In addition to nonpayment for services, most of my interlocutors shared stories of extortion, death threats, and overt violence to degrees far beyond what one might entrust to the courts to adjudicate.

Furthermore, several RSIEA graduates who work in environmental architecture, but whose employers are large builder firms, were unwilling to speak with me on an ethnographic record at all. All apologized, with some expressing a sincere wish that they could share their experiences with me. Yet even the appearance of sharing information about that builder's designs or the processes they followed in order to have them built could cost these architects their jobs; this was a risk they could not, neither would I ask them to, take. Any promise of anonymity was automatically insufficient, and the ambient sense of fear and danger that accompanied most discussions of corruption in Mumbai's urban development omnipresent.

Aditya assured me that in his experience the government tendering process was as fraught as the processes others described in the private sector. An inevitable choreography was predetermined, and it ensured a substandard buildingscape. He explained:

With the government buildings the tender process is long and complicated. Let's say that you and I are contractors, and a tender is floated by a government body at the cost they expect for the project. Say 50 crore is the cost of the project. You are a contractor and I am a contractor; you will charge 51 crore, and if I quote 49, I will get the contract irrespective of my actual calculations of the cost. No matter what background I am from, if I can quote it for 49 I will get the project. . . . Now, that same firm has actually calculated a total cost for a project of 35 crore. They will stretch with the unskilled workers, maybe they will take labor from a place where they don't have knowledge of construction, you know, they just make it work without the costs. . . . Why? Because if you want to get a government contract, then the practice is like that: you lower the quote to get the job. This is not how it should be. If you want good buildings it should be a good contractor. It should be good infrastructure. How can we talk about environmental *anything* when this is the process?

Inevitably this reliable choreography foreclosed any latitude for an architect to make any kind of intervention, whether based on good design intentions or not. Aditya continued,

> As the contractor I quote that low, but I have in my mind some way to make up that money. It's like a lump sum contract. Now I come as the architect and I look at the drawings, and I see things that should be changed or improved. This is a normal thing for architects—I am trying to improve on some things. But if I do that, the contractor will say, "this is not under the contract," and they will demand extra money as devaluation cost.

When I observed that this would prevent any architect from revising the design to make it more ecologically sound, Aditya smiled and paused. "In fact," he said, "that will cut my amount." "But this means that you can't make any suggestions, environmental or otherwise, without incurring the cost yourself?" I asked, realizing that the personal material stakes of this kind of intervention were rarely mentioned in the entreaties to action that closed the most inspiring of RSIEA lectures. "Well," he answered:

> If I find mistakes in the drawings, and I try to change something, and my boss agrees, then there are just a few sides to convince—the client side, the government side, and the project manager. You discuss it. But just the other day I went to Goa (to check on a project we are doing there) and I asked (my boss) if we can make some very important changes. These were for safety, not even for green aspects. But he said no. So here I am; I came back to Mumbai and I find today that even my structural engineer has made mistakes. More mistakes! But I cannot change these things without cost and without convincing all the others. And they are not interested.

Aditya's experience with exclusively high-level residential development in Mumbai left him adamant that there is no scope for practicing anything that even resembles RSIEA's version of good design—let alone space for reform in the urban development process—in Mumbai itself. Others, however, like Suhasini, Prisha, and Darius saw recent work stoppages associated with the municipal commissioner's tightened permissions enforcement, discussed earlier, as reason for optimism. "I think it is already changing," Suhasini said; "this new commissioner looks good in principle, and I think he has some integrity." What was notable, regardless of whether powerful structures in urban development were changing or fixed, was the degree of powerlessness that each architect described when it came to catalyzing that change. Each environmental architect I interviewed outlined an ecology in practice that seemed in every way circumscribed by a "system." Changes to that system were almost never considered to be within the architect's purview.

Yet, the system *was* changing. Darius pointed to a significant shift in old norms, which he called "loopholes," that had previously allowed builders to create larger flats than Floor Space Index regulations would allow:

The old rule was that when you design a lift, what is in front of the lift is free of FSI. So builders would have architects design one lift that would open this way, and another that would open (the opposite) way. So in the municipal drawings they are shown that way: all correct. But in our drawings, the space at the back, which would be 200–300 square feet, would go into the flat. This was also common practice with car lifts. You had a car lift, and builders would say they are providing parking spots in the air. . . . That's one thousand square feet, and you were getting that FSI free. Technically it wasn't illegal . . . but then of course people didn't use that as a car parking spot. They used it as an extra bedroom or extra living room, a den. These are the kind of loopholes that municipal architects informed us about. This is something they could do, and we exploited that.

Again, the agent of change was not the architect; it was the new municipal commissioner, a state actor who was reforming the structural choreography. The dynamics of agency and power had not changed; architects may be able to do things differently, but they were still constrained to Darius' "system." He continued:

But these days (with the new commissioner), those very same things are not allowed (and so they are causing work stoppages). . . . Just another example: earlier, balconies were all free of FSI. The commissioner changed that too. We were designing two thousand square foot balconies, one thousand square foot balconies. . . . I mean you don't have *flats* that size. And those were all free of FSI. So in a, say, four thousand square foot flat, we were giving you a two thousand square foot balcony. This was completely free; that space wasn't counted in FSI. The builder gets one hundred percent on that because it's not taxed. So the development people see it as a win win-win, but of course the BMC that was losing out on this money, and the city was not able to control the environmental damage or other costs. So what the new commissioner has done is to simply say that if you make a balcony, it's considered in FSI. Everything that was free earlier, you'll have to pay for now. It's amazing what a complete change this simple thing has made.

Despite the overwhelming experiential evidence that an established political economy of development—however it may change in terms of specific regulations—deeply constrained environmental architects' attempts to practice good design, I nevertheless heard repeated assurances that eventually the scope for practicing good design in Mumbai would change. Amrit, for example, was unapologetic in his optimism, and always quick to return our conversation to the possibility of agentive, collective action through architecture itself. Despite what seemed to be an endless set of stories of architects' powerlessness in practice, the source of this optimism derived from an aspirational capacity to act. To one of my more skeptical questions about what appeared to be the impossibility of environmental architecture, Amrit simply said:

Look, I'm very optimistic! But I'm realistic, too. I do think architects will need to come as a force if we want to change things. *We* should enforce certain things on the builders . . . not builders enforcing certain things on the architects. Of course it is possible, and in fact I think it is inevitable. It's a matter of time, and those of us who are willing, we need to keep fighting for this.

Such comments resonated with the almost unshakable faith in the inevitability of significant transformation—and thus significant new opportunities to practice good design—that opened the discussion in this chapter.

Suhasini told me that her optimism came from a diehard idealism, a personal characteristic in which she took great pride. "It's always a struggle," she said, "but change is constantly happening, all around the world."

> Everyone in my office is highly idealistic. We believe change can happen because we know it is happening. Look, even these little things in Mumbai. In my office, we all come on foot or cycle; no one comes by car . . . and I'm happy that (I work with) a class of people who don't see the cycle as a poor man's vehicle. It's a small, small thing but it is uplifting. So I know the bigger picture *will* change, but it will take time. We need a whole paradigm shift. It will take time. I know from my Rachana Sansad class there are many people who want this! I know people of my generation who want to cycle. See, you have to start to make a trend and soon it is happening.

Prisha nodded, adding, "I still think there's a lot of scope in Bombay. There are people who are actually practicing environmental architecture, and they are trying to push the limits. Sometimes they are even doing it. It's a slow process. But we feel it."

When I asked Suhasini if she believed that change would depend on idealists, she was quick to say, "No. In fact it's the common person who has to make this happen." She explained:

> Many people travel abroad these days, you know, the first-timers who go abroad. They come back saying it's so clean—the air is fresh and they have parks and good transport—in other countries. So I say, then let's do it here! But there are always those people who say no, it's not possible in India. Why is it not possible in India? The main thing is cultural: you have to *believe* it's possible. I think part of why there is no willpower is because there's corruption everywhere. You can just go to any MCGM office and show them a very good design, and they are still just dismissing you, talking about idealism, come to real life, it's not possible. But that's generational and it is moving out. I think eventually—hopefully it will move out. I am very positive. Think of how Mumbai was ten years ago or fifteen years ago, and what we are, where we are today, it's good. Another fifteen years and there will be a lot of change. Maybe not a sustainable city, but we will be able to *be* environmental architects.

I persisted, asking if inevitably environmental change would have to be catalyzed in the realm of politics. Perhaps it's true, I suggested, following the exchange at the start of this chapter, that policy changes, combined with enforcement, are the only ways to solidify a new direction for the urban development future of Mumbai.

Suhasini agreed, but added that it will also depend on architects. "I don't have to be a minister to make lasting change," she said, continuing:

> All over the world, people are changing things. It is environment, but it is also political. Look at the climate change, look at urbanization, look at all the political demonstrations. Things are changing. But if you look at (Mumbai) right now, in 2012, you wonder, what did we learn from our Mumbai floods? ... Nobody is really trying out new ideas (at large) scales yet ... so it's like if you have a tumor and you're just eating medicine, but not removing the tumor. In Mumbai we reclaimed (urban land) in the wrong places and everything is flat, so all the river drainage is gone. So (the floods are) just how nature acts. But see, it cannot continue like this. People will demand change, and nature will force it. The politics are just reflecting this.

I continued my line of logic, asking, "So architects are also powerful in this?"

> Yes. Doing environmental architecture *is* a political step. And it's also right now a huge risk. But if you do take the risk, then ten years down the line when the rains come like that again, you have (a different outcome). My father is of that older generation, and so of course he (is skeptical), saying that already Bombay is too small and land value is so high, so you can never take land away by changing the (reclamation patterns). I'm not saying make Mumbai seven islands again ... it's not possible. But you know, by thinking this way you actually see that in Bombay, if you organize it correctly—environmentally—there *is* a lot of space. You will even have open spaces! So if you organize it well, and you have smart design, it will happen. We have to think differently than my father, than the officials in the MCGM. We have to, you know Ian McHarg? Like, design with nature.

Suhasini's dual faith in a generational shift and the capacity of that generation to both imagine and organize urban space differently was pervasive across many interviews—often to an extent that defied easy analysis, particularly in light of equally adamant descriptions of an existing system that not only constrained architects' choices and parameters for action, but was also riddled with risks, dangerous facets, and unscrupulous practitioners.

Environmental architects often framed the factors that enabled or constrained good design practice in very personal terms, through one's material capacity, consciousness of scale, or beliefs about the impact of individual consumption patterns. Consider the contrast between Aditya's discussion of the financial circumstances that constrain the kinds of architecture he can practice, for instance, and Kalpana's relative freedom to make choices about the firm that she runs. In the latter case, the constraints had more to do with the scale at which one assessed a design's impact; in the former, constraints began with the essential need to earn a specific salary. Aditya explained:

> After graduation, after three years, I have not found any firm where I can practice what I learned (at RSIEA). Yes, there are firms in environmental architecture, but they cannot pay that much, and I have to keep in mind financial stability. I get a

certain amount in my current office. Here I can do small, small things. But pure environmental architecture, if I go there they would pay me half as much. It's not sustainable financially for me . . . so, I simply cannot work full time in the environment field. I would love to, but it's not viable. I just bought my flat. I have my dues to pay for that, and if that is not feasible, then I am an architect without shelter! . . . How can I work for half the pay?[20]

In contrast, Kalpana offered:

I have my own firm; there are two of us. As and when we need people we hire on a contract basis. What I'm doing right now is essentially more interior work and small homes outside of Bombay. A couple of commercial complexes also. I am able to implement most of what I learned (at RSIEA), in terms of (using green) materials, energy systems, and general design principles, because I share that kind of a rapport with my clients. I sit down and I talk to them about what they want, and I can say, "You know, why not build it this way? Even if it costs you a little bit more . . . And they are willing to listen. We got an offer to design a school, for example, and they came with these pictures saying, "This is what we want, this is what we want, this is what we want." And I said, "You know, okay. But why don't you look at it in another way? So I showed them some pictures of the school in Auroville, and I said, "You know, this is as good, if not better, (than the pictures you have). And they realized, like, "Oh yeah, this is much better."[21]

In the contrast between these two responses, the constraints of individual architects' own positionality within the political economic system—not just as professionals, but also as individuals, were made very clear.

Yet beyond the question of one's own class position and position within one's firm, several issues also emerged in architects' discussions of the factors that enabled or constrained good design practice. For many, RSIEA's emphasis on good design's conceptual dimensions came at the expense of what they called more "practical," skill building classes. This left environmental architects relatively underprepared to implement what they had learned. On one example, for instance, Kalpana, Palavi, and Arash, characterized this issue in terms of ideology and strategy:

Students discussed at length the ways that the (RSIEA curriculum) is "all ideology," teaching very few skills that they called "technical" and "practical." Arash asked why the curriculum doesn't split into two parts, in which the first year is "ideology" and the second is devoted to "practical" concerns. He argued that as he was on the verge of graduation, he felt he should be able to design a green building. He suggested that the Design Studio final project—the resort at Pali—could have included the stipulation that it be a platinum rated building, for instance. Arash emphasized a need for more classes or projects that focus on hands-on calculations and "thumb rules." Otherwise, he said, environmental architecture remains intuitive, even as they are graduating.[22]

At the same time, architects emphasized how central that same "ideological" dimension was to their sense of professional identity as an environmental architect.

Even for Siddharth, who, the reader will recall claimed to have no particular interest in environmental architecture, told me:

> I was never interested in *environmental* architecture, frankly . . . but the course
> helped me a lot. It completely changed the way I think about *architecture.* When I
> used to plan buildings before, a building was a footprint on the ground. But when I
> learned environmental architecture, and that you should design a minimalistic foot-
> print that is integrated with the ground . . . it forced me to always think in a creative,
> different way. You think in a way that doesn't endanger nature. And it makes the
> overall design so much more interesting. I would never want to go back to the way
> I used to practice.[23]

Architects also described the ways that the philosophical dimensions of RSIEA
training left an imprint that reached beyond their professional practice. Suhasini,
for instance, told me that the moral ecological aspects of learning and embracing
good design had transformed her very sense of self:

> I needed RSIEA to show me a path. Maybe if I had (just learned the technical aspects
> of environmental architecture) on my own, I wouldn't know what is right or wrong, I
> would just go whatever way. Definitely I could have done without it from the point of
> view of technical skills, but yes, it did help. It changed me as a *person!*[24]

Similarly, Palavi identified critical engagement with environmental design as a
lasting, if not directly "practical," or even enabling, skill:

> The course makes you think and question things. It makes you question the current
> practices, the current materials, but then even everything that you do. . . . You ques-
> tion and you also think about what is the right process to assess things. What is the
> right process to know what is correct, environmentally? The program guides you to
> develop that thinking and to feel you have grounding for your ideas. And you know
> you are doing something good.[25]

Even Aditya, whose personal financial constraints so restricted his capacity to
practice environmental architecture after graduation, was enthusiastic about the
positive impact of the RSIEA experience on his life and his work. During one of
our conversations, we spent most of an hour discussing the ways that he actively
promoted RSIEA to his colleagues and friends. True to his own economic situation, he described the benefits of his training through a cost and benefit equa-
tion, settling on the unquantifiable dimension of an experience that "guides your
conscience":

> The course is two lakhs for two years; IRs 40,000 per semester, plus college entry fees
> of IRs 10,000 or something like that for six months, so for two years it comes to two
> lakhs plus the study tours. This is not easy to afford. I went to all the study tours and
> they are also an expense. And now, many of my friends are asking, "Aditya is it really
> worth it? Should I join?" I say, "Yes!" Because this degree is only two years, and yes,
> you are paying, but you are also earning at the same time. You can keep your job,

and be in your office during the week, so you keep earning. . . . But this course has its own importance. It guides your conscience. I would do it over again and again. If you're really interested you can learn anywhere, and with RSIEA I learned the big picture of green.

Our conversation ended in laughter, as Aditya assured me that, having studied at RSIEA, "I know what environmental architecture is, so I am also very good to tell you what it is *not* in Mumbai."[26]

THE ENVIRONMENTAL ARCHITECT AS INTEGRATED SUBJECT

Newly conversant in the techniques of good design, RSIEA architects faced the challenge of ecology in practice. Through their accounts, we glean a sense of urban material and economic development in Mumbai, and how practitioners discerned, experienced, and engaged its organizing systems and power structures. Social categories like the state, the builder, and the architect in its many guises organized narratives of purposeful and dynamic actors operating in an otherwise "chaotic" process called urban development. Each architect described his own position within that process, placing the self, and the category of the "environmental architect," in strategic relationship to the figures and forces understood to limit or shape good design.

Graduates' narratives also conveyed individual and collective logics of when, and how best, to challenge existing structures with an eye toward transformation. Their retellings of ecology in practice trace the simultaneous scales and forms of transformations already underway—from new officers in specific municipal positions to global protest actions—and patterns of power, economic incentive, and established processes that limit, if not fully foreclose, any chance to enact good design. In conversations across a wide range of RSIEA graduates, from the newly finished to the seasoned practitioner, the tension between finding an evidentiary basis for good design aspirations and describing, as Darius called it, "the system," were always present. If that tension was to break, it seemed, it would be the environment itself, rather than environmental architects, who would force it: ever-more untenable conditions of resource scarcity, pollution, human depravation, and suffering seemed poised to open the future to good design. As Suhasini declared with both confidence and optimism, "people will demand it, but nature will force it."

As professionals, RSIEA architects' newly cultivated sensitivity to biophysical processes and the functionality of urban ecosystems allowed what they considered to be a unique, and often superior and more complete, perspective on Mumbai's urban past, present, and future. Suhasini's reference to the Mumbai floods is a useful example. She locates their cause deep in urban land reclamation history, but in a final assessment concludes that contemporary land distribution, land value, and political economy will prevent any reversal of resulting drainage patterns. Nevertheless, she proudly proclaims herself an "idealist," emphasizing her confidence in the inevitability of significant change, and perhaps, ultimately, "reorganizing" Mumbai in ways that will recognize the legacy of those historical drainage patterns. For her, a generational turnover in municipal and other state agencies will accelerate change, making today's very circumscribed environmental architecture interventions merely temporary.

If the agents of change formed one set of important concerns that emerged as architects sought to enact ecology in practice, then the pace of change formed another. The capacity to understand the origins of contemporary environmental conditions and events, such as the Mumbai floods, in terms of longtime patterns of land reclamation and morphological modification left RSIEA architects equipped with an intellectual basis for their expectation of more, and more intensive, catastrophic ecological events. This is critical for understanding a pervasive, underlying expectation not only of massive eco-social transformation, but also of the scope for their own opportunities to operationalize what they had learned. By studying the environment as an integrated subject, and environmental design as good design, environmental architects might themselves be understood for their own integrated subjectivity: although ecology in practice was largely aspirational in the present, it was not only prudent, but prescient. It was through their own integrated subjectivity that RSIEA environmental architects anticipated an essential place in the Mumbai to come.

Meanwhile, everyday strategies to resolve the impasse between good design training and the (current) realm of the possible were in part driven by pure pragmatism, as in the case of Darius, who said, "you accept it; you move on," or Aditya's logic of necessary financial and professional survival. These strategies were not, as Aditya joked, "anti-environmental" architecture; they were simply grounded in the dual belief that wholly necessary practices were, in the present, conditioned through the actions of other, more powerful urban development agents. Rather than assume the idealized role of an activist, RSIEA architects held their commitment to good design in service of a vocation in waiting. It was the environment itself that provided legitimate reassurance that the wait was worthwhile.

Likewise, pride and personal integrity grounded in RSIEA's brand of moral ecology implied a critical mindset relative to urban development norms in Mumbai, and the firm belief that environmental architects could maintain their integrity despite their embeddedness in a system rife with opaque norms. Nearly every architect described herself as able to remain separate and autonomous from the ubiquitous layers of illegality and corruption. Their experiences of a "messy" present recalled the comments shared earlier by a member of the RSIEA planning faculty. However pervasive the problem, that professor assumed a standpoint guarded by the possibility of preserving an objective, neutral stance that enabled moral insulation. This she held, despite a city in which:

> The kind of mess they have created now even the builders can't help it. Even the pan-walla is demanding with an old building that is to be demolished . . . the illegal tenant who (occupies it) has a goonda . . . It's very easy to tell that builders have become the leeches who are taking the blood from the city, but now the small fries are also sucking blood from the builders. In huge numbers. The other day the (municipal) commissioner was saying that every small and big citizen of Mumbai has become a blackmailer, because he is in a position to take advantage of the legal system . . . (Even) the middle class people are (doing this). It's everywhere, at every level and completely normal.

Such narratives of neutrality ultimately crystallized in a moral ecological mode of belonging, one that mapped to the simultaneously technical and conceptual aspects of good design, but that completed its contours with repeated appeals to consciousness. Environmental architecture as good design—consciousness, critical mindset, and specific techniques—was a lens for assessing present eco-social dysfunctionality and working in anticipation of future transformation. It was the means for organizing one's understanding of the relationship between individual politics and commitments, professional choices and imperatives, and the sometimes Sisyphean task of environmental changemaking writ large. Good design provided a metric for personal integrity, bureaucratic transparency, and indictments of categories like builders, state agents, and municipal architects.

Considered together, these narratives of ecology in practice allow an extended consideration of the engaged, practical fate of RSIEA's good design training formulation. They demonstrate that it is not only too simplistic, but in most cases simply inaccurate, to suggest that architects graduate from RSIEA with delusions of endless and transformative agentive capacity. It is similarly shortsighted to assume that aspirations formed in training simply vanished over time or through the wear of experience in the complex of forces and structures that shape urban built form development in Mumbai. Many current and graduating students, as well as RSIEA graduates across the history of the program, described profound and resounding certainty in the inevitability—somewhere in the near future—of urban environmental improvement in Mumbai. Ideas about when, precisely, it would come about, and the exact scenario that would catalyze it, differed dramatically, though; so, too, did notions of what would become of Mumbai's built form environment, its biophysical environment, and its social worlds in the interim. The moral ecology of good design, resilient appeal of green expertise, and assured anticipation that the environment itself would set its stage, was almost ubiquitous. This was the basis of ecology in practice.

8

Soldiering Sustainability

Expectations of ecological decline in cities are well rehearsed, and nearly always framed against a backdrop of unprecedented growth, unprecedented climatic conditions, and unprecedented movements of those displaced by ever more precarious environmental and geopolitical circumstances. The Anthropocene looms large and ominous, and its biophysical and social realities embolden anxious responses.

This book set out to understand a collective social response to the urban present, and the urban future. It traced how ideas of ecology and nature were integrated into a specific architectural modality. The geographic and historical setting within which this modality was embedded—contemporary Mumbai—is undoubtedly unique, but the central questions at hand resurface across urban contexts in the peculiar, uncertain era called the Anthropocene. How, I asked, would agents craft a social mission to transform the built form of Mumbai, and how would ideas of ecology galvanize it? What kind of sociality would adherents to that mission forge and inhabit? Once established as an environmental affinity group, could environmental architects actually make the kind of change they were now collectively equipped to envision?

I addressed these questions in a historical moment when Mumbai is riddled with seemingly intractable environmental and social problems, yet also buoyed by robust economic growth, narratives of global ascendance, and the bold confidence captured in slogans like, "Consider it Developed." This tension framed the research in this book, and cast into the foreground the ways it is lived in everyday social and professional life. As my interlocutors at RSIEA described their aspirations to understand and practice good design, they demonstrated the force of a collective moral ecology, one that conditioned striking—often seemingly impossible—spaces of imagined possibility. These emerged repeatedly, in spite of an endless array of bureaucratic, political, and economic obstacles to operationalizing good design. As I have shown, the confidence that RSIEA students and graduates espoused cannot be fully captured, or simply dismissed, as bourgeois delusion or rehashed technological optimism. Although both privileged positionality and a context of ever-evolving technologies were certainly facets of the RSIEA experience, architects' repeatedly reflexive posture toward both, and their commitment to specifically rendered logics of equity, justice, and more-than-human nature challenge us to think beyond more conventional political ecological analyses. The force of their shared moral ecology played an undeniable role in fostering RSIEA architects' collective refusal to imagine the future of Mumbai within the political economic and material conditions that characterize its present.

At the same time, such confident aspirational politics are not new to environmental activism or to urban design; neither are they unique to the broader tradition of social uplift through environmental politics present in the many forms of postcolonial environmental action across India.[1] A host of examples might be found in arenas of indigenous knowledge or tribal land rights, for instance, or through historical figures like Anil Agarwal and his Center for Science and the Environment; these often made an explicit point of amplifying the positive developments and hopeful ideas that would energize aspirations for ecological transformation.[2] Many such figures populate the history of Indian environmentalism, urban planning and design, and social justice work.[3] The rhetorical promise of the Jawaharlal Nehru National Urban Renewal Mission to improve city life and infrastructure, along with its program of Basic Services to the Urban Poor, provide other examples; so, too does the Atal Mission for Urban Renewal. Although riddled with political complexity and often vigorously critiqued, such initiatives provided professionals of a previous era with concrete policy rubrics that sketched the form of the possible, and, perhaps, energized their adherents in ways analogous to the case in this book.

Yet the aspirational politics formed and reinforced through RSIEA's moral ecology of good design would be misread if we were to regard them as galvanized exclusively in nationalist or regional political registers, or exclusively anchored to local and regional scales. For RSIEA architects, nested global, national, and regional circumstances set the stage for an *inevitable* rise of the environmental architect, however dormant or constrained she may be in the immediate present. Global developments as varied as the rise of green building certification systems worldwide, the proliferation of comparative mechanisms that ranked world cities according to environmental conditions and achievements, and impactful sociopolitical movements that ranged from Occupy Wall Street to Occupy Central, stood as evidence of global transformations that gestured increasingly outward, and thus political economic spheres that, however locally conditioned and embedded, had political ecological logics that ensured and reinforced their legitimacy at every scale. The so-called inconvenient truth of global climate change itself, perhaps, stood as the ultimate condition that separated the RSIEA mission of the past from its moral ecology of good design in the present.

Throughout the book, I traced how a distinctive sociality—framed as an environmental affinity—was produced. It combined RSIEA's version of green expertise with a post-training commitment to ecology in practice, and drew from a wide set of references that, through RSIEA training, came not only to be shared, but also to connect a grounded sense of good design to a much larger regional and global environmental sensibility. Good design, in the sense of its collective sociality, came to demonstrate how the specific work of urban ecology—that knowledge/practice hybrid bridging ecosystem ecology, social process, and material form—proceeds in social life.[4] The work of urban ecology enfolded both an integrated subject *and* the integrated subjectivity of the social agents who espoused it.

The RSIEA case also underscores the peculiar temporality of contemporary urban sustainability as it is lived in social life: actual good design was always deferred, and yet ever more urgent. I argued that this temporality was central to environmental architects' aspirational politics; the temporality of its own fluorescence depended fundamentally on dramatic, if not catastrophic, ecological and political shifts. RSIEA-trained environmental architects viewed theirs as a vocation in waiting precisely *because* certain environmental and political changes would inevitably ensure the need for their skills. The only uncertainty was whether the primary catalyst would be the environment or human politics; whether the path was opened by nature, human society, or both, future practitioners of good design stood at the ready.

Yet to stand at the ready with bold confidence while also embedded in the everyday structural realities of Mumbai's urban development involved constant, undeniable compromise. And thus emerges the second unresolved tension in the book: despite its socially vital life as aspiration, in practice, good design was inevitably dormant. Even standing at-the-ready, its enactment seemed always almost fully dependent on external political economic and ecological activation. We might therefore dismiss good design's adherents as politically benign at best, antipolitical at worst.

Yet I suggest that we risk a great deal if we are to dismiss the complex social life of good design as simply ineffective or benign. In part, this argument returns to the peculiar temporality of urban sustainability itself, but it also underscores the very specific ways that conceptualizations of "consciousness," and explicit formulations—however problematic—of a rhetorically secular and inclusive notion of Indian identity point to something more complex. The environmental affinity forged at RSIEA and lived as a vocation in waiting illuminates a steadily growing social arena in which a shared moral ecology repeatedly reinforced an ecopolitical mission. In the process, it reproduced and sustained the essential resolve that ensured that the wait was not in vain. The vocation in waiting, I contend, was a space that nurtured an urban environmental politics that, though dormant in the present, may at any moment find a force to ignite it.

The Mumbai context, of course, carried with it significant place-specific political stakes, even as the explicit contours of the contemporary political economy of urban development in Mumbai remained unnamed and un-discussed in the context of RSIEA training. Grossly uneven relations of power and access to resources are a clear facet of the city's everyday life, and to ignore those circumstances is to assume a complicit position within them. While it would be inaccurate to interpret these silences as evidence that RSIEA students and faculty did not genuinely care about putting good design into practice, they do invite us to think carefully about how and when environmental architects configured the temporal calculus of socioenvironmental transformation and socioenvironmental justice. Good design carried with it clear moral imperatives, but it also enabled a logic of deferral that allowed architects to make repeated social and environmental compromises in the present without violating the eventual mission.

The shifting scales of reference so central to this calculus—sometimes focused on the neighborhood, at other times, the city, and at yet others, entire watersheds, ecosystems, or non-human species populations—allowed a constant slippage between articulating various costs, and rescaling logics of benefits, that could be derived or would accrue. Open, green, or vegetated city spaces, more efficient energy use, or cleaner air and water might be valued as "public" goods with intrinsic capacities to remedy urban unsustainability, for instance, even as gross inequalities might persist between social groups when it came to which groups might enjoy direct access to those spaces and their benefits. Such formulations of eco-social costs and benefits was most visible in the case of the Doongerwadi Forest, evaluated as it was for its broader array of non-human natural attributes and processes, but open only to an highly exclusive human public for any direct use or experience.

Scalar logics of good design introduced yet another clear tension, one particularly visible during study tours. Many of its *sources* for ecological ideas, values, and strategies mapped to population-sparse, space-rich contexts; nearly all were sited in contexts *other* than Mumbai.[5] The question of precisely how, for example, a decentralized water management system observed in Auroville might be applied to a design brief in Mumbai was left unresolved, leaving such questions of how to scale up, urbanize, and otherwise modify various lessons to fit the Mumbai context unsatisfied. Although students and faculty were fully aware of these discrepancies, few discussions took up the direct question of their Mumbai-relevant analogue. The contents of the toolbox, in this sense, often stood remote from their intended sites of application and relevance.

But perhaps most critically, good design's expansive, multiscalar calculus of environmental justice included the benefits that would accrue to non-human nature, with the effect of recasting logics of equity and the social good as logics of equity and the socio*natural* good. This further legitimized, as in the Doongerwadi case, exclusive, controlled access to the forest in a city otherwise starved for open spaces. Again, this reinforced environmental architecture as almost automatically noble: even in its deferred state, good design "counted" as a mode of doing good.

We began, and we end, then, with a city whose future material form is still largely unbuilt, and still in-the-making. It takes shape in real time, however rapidly and however divergent its path from the influence of the good design that RSIEA architects espoused. Demands for open spaces, projections and creative renderings of a future city mosaic of built forms and urban natures, and stark socioeconomic inequalities all punctuate the triumphant ascent of this city at the economic heart of "India Rising." The more general consolidation of political power on India's nationalist right has brought renewed international attention to India's social and political economic future, as well as to its environmental one. Where and how the temporality, aspirational politics, and moral ecology of RSIEA environmental architecture will fit in the Mumbai of the future is a chapter that continues to write itself in real time. The ultimate resilience of good design as ecology in practice remains to be traced and observed, but if its proponents are correct, it may be the environment itself that remakes Mumbai's political stage, perhaps sooner rather than later.

Green experts, their publics, their spectacles, and their hybrid knowledge forms, all provide guidance for reading the city of the present and anticipating the city of the future. But they also caution us to disaggregate them, noting the difference between green knowledge forms and their in-practice social lives, and the temporalities through which they galvanize moral and political force. While green knowledge forms may foreground the integrated subject, their lives as ecology in practice demand more careful attention to the power relations, aspirational politics, and enduring social structures that organize the moments and contexts within which they may be operationalized. However dormant or deferred, RSIEA's modality of good design gave the urban future a social life that could be both lived in the present and practiced, as anticipated, in the future.

In its dual arenas of training and practice, RSIEA environmental architecture challenges us to move beyond conventional political ecology analytics in ways that can more fully engage more-than-human agency, more-than-human exclusions in our analytical calculus of equity, and the aspirational politics that characterize the socialities of human agency-in-waiting. It challenges us to reconsider the presumptive authority and agentive power of the green expert—the so-called "soldier of sustainability" who understands, as students were assured, what others do not. In its guise as good design, RSIEA's environmental architecture reminds us that ecology in human, agentive practice depended in large measure on the ways that practice was made meaningful; it challenges us to forge an analytical place for the political purchase of agency deferred. A vocation in waiting, I have argued here, constitutes an arena of politics worthy of attention—one suspended in the social structures of the present but unquestionably confident in the inevitability of structural transformation through the agentive capacity, quite possibly, of non-human nature itself.

I deliberately left the starkest reminders of the power of existing urban development in Mumbai to the end of this book. My aim was to underline that in the case of good design, if we were to measure effectiveness by real-time implementation, there may be no book to write, no ecology in practice to explore. If it can't be *practiced*, after all, how can it have social or environmental value? Indeed, good design as practice was heavily constrained and usually curtailed; it was nearly always foreclosed within political economic structures and bureaucratic orders that showed only passing indications of any transformation at all. If we were to start and end the analysis with the work of the present, we might rightly look to arenas of finance, politics, real estate, and governance for the real—and only—story of environmental architecture (and its absence) in Mumbai.

Yet the sociality of building green was real, powerful, and perhaps even profoundly political, despite the material absence of much that "counts" as green building. To confine our understanding of RSIEA environmental architecture only to its evidence in the material built cityscape is to miss its social, political, and ecological point. Good design's feasibility depended on far more complex metrics, expectations, and temporalities, to say nothing of a more expansive understanding of aspirational politics *as* politics. As Arjun Appadurai has argued, forms of hope and anticipation like those which organized good design are always in tension with aspirations configured by "the dreamwork of industrial modernity, and its magical, spiritual, and utopian horizon, in which all that is solid melts into money."[6] Indeed, the city unbuilt is also the speculative city, but the possible future within which to imagine fabulous profits and solidified asymmetries of power and wealth may also collide with, and give way to, a very different, and yet perhaps equally plausible possible future marked by transformed environmental conditions, transformed politics, and an urban form ripe for the good design practices of RSIEA's environmental architects.[7]

Our challenge is to consider together, and to understand in tandem, both the speculative dreamwork of capitalism and the dreamwork of good design.[8] The latter articulates a regime of value more expansive than capitalist calculations can capture, yet positions itself in the present within the deceptive arena of bourgeois, professional practice. It seeks simultaneously to do well, and, eventually, expects to be positioned to do good. It espouses a moral disposition that works to embody practices of ethical engagement, and it expects of its labor a materiality that only multiplies its positive ecological *and* social effects.[9]

Recent scholarship on new materialisms has suggested that a truly ecological study of the dreamwork of capitalism would give close attention to the sometimes profound and unforeseen ways that materials may be regarded by their users as things with the potential to bind human beings and the non-human, biophysical world in new ways. After all, it was the promise of good design itself—embodied in form and declarative of social aspiration—that bound together the moral disposition that made ecology in practice thinkable in the Mumbai of the present.[10] This is not a trivial fact; it suggests social actors with clear belief in the birth of alternative political economic spaces to be born inside the very latticework that industrial capitalism continually reweaves. In the present case, the catalyst for that alternative political economic space may be non-human nature itself, however evasive it remains of our usual analytical toolbox.

If a broader political economic critique of global environmental urbanization—indeed, planetary urbanization—traces cityscapes of ever intensified vulnerability and suffering, it also quite notably returns ethnographic facts that trace spaces of astonishing aspirational hope. Once opened and activated, they remind us not only of capitalist dreams of future value, but also of the more-than-capitalist—indeed, more-than-human dreams of a different, possible city. They challenge us to take seriously the deployment of shared environmental affinities as a conscious mode of social inclusion—even in historical moments when the very symbolics on which they draw are otherwise heavily marked by their violent promise to exclude.

Even as IndiaBulls fragmented into scandal, Mumbai's new development plan suffered repeated delay and controversy, and enthusiasm for reimagining Mumbai notably waned, RSIEA's environmental architects remained. Their numbers grew, and their affinities strengthened. They may even, in fact, be stronger than ever in their own generational logic of imminent and totalizing change. The fundamental source of their inspiration was neither the fate of the development plan nor the satisfaction of putting their newly gleaned green expertise into immediate practice. It lived on, instead, within a sociality of environmental affinity that emboldened collective confidence in inevitable change—a confidence more robust and meaningful today than it was at this writing.

PREFACE

1. Prakash, 2010, Pp.26–27.

2. A rich literature explores the architectural and design dynamics of colonial governance in the making of late nineteenth and early twentieth century Bombay. While work
like Metcalf's *An Imperial Vision* argued that there was a distinct colonial intention to foreground notions of traditional society in its architecture, and to link these notions to fostering acceptance of European ideas of progress, more recent work—notably Chopra's *A Joint
Enterprise: Indian Elites and the Making of British Bombay*—shows how the city was forged
as a dynamic and shared endeavor. Chopra rejects a historical model that assumes that
Bombay was made only by its British rulers, showing instead that both colonial and Indian
elites forged the city in negotiated dialogue. Chopra points to specific buildings, their plans
and their styles, as well as to specific figures from architecture and planning, to demonstrate
the hybrid qualities of the "joint enterprise" of making Bombay. Most often, the Indian
elites in question are Parsis. Additional work along this continuum includes Evenson's *The
Indian Metropolis,* Dossal's *Imperial Designs and Indian Realities,* and Kidambi's (2007) expansive *The Making of an Indian Metropolis: Colonial Governance and Public Culture in
Bombay, 1890–1920.* Considered together, these works demonstrate a scholarly trajectory
that has come to appreciate the negotiated qualities of urbanization in Mumbai and the
complex interplay of power and social positionality that produce any city's material forms
over time. For the making of contemporary historical narrative in Mumbai, see in addition
Mehrotra, 2004.

3. See for example Patankar et al. 2010.

1. CITY ASCENDING, CITY IMPLODING

1. For a review of the intersection of urban planning and utopian thinking across a range of historical cases for South Asia, see Srinivas 2015.

2. See Anand 2017. In Anand and Rademacher (2011, 5–7), we discussed how the past two decades of economic liberalization and globalization, in combination with other nation-level reforms, have produced significant economic and ideological transformation in India. Foreign-capital driven speculative investment in newly opened urban real estate markets led some observers to describe economic change in India as "casino capitalism" (Nijman 2000). But even as Mumbai has been regarded as a city awash with cash, commerce, and consumption (Appadurai 2000), it is also a global icon for discussions of urban informality, inadequate housing, and the patterns of neoliberalism, capitalism, politics, and gentrification that occupy policymakers, scholars, and shape the lived experience of informality and marginality. As we wrote in Anand and Rademacher (2011, 5–7), over half of Mumbai's population lives in settlements that occupy only eight percent of the city's area (McFarlane 2008); eighty percent of these live in homes of less than 100 square feet (Sanyal and Mukhija 2001). Over the last century, the Maharashtra state and Mumbai municipal governments have addressed inadequate housing through simultaneous strategies of accommodation, regulation, and demolition (Chatterji and Mehta 2007). See also Roy 2009; Doshi 2013; Weinstein 2014.

3. See note 2 regarding histories of colonial urbanization processes in Bombay.

4. See, for example, Prakash 2010; Hansen and Varkaaik 2009; Chalana 2010; Weinstein 2008; Rao 2011.

5. See Davis 2006.

6. See, for example, McKinsey Global Institute 2010 and Burdett 2007.

7. The global popular press included accounts such as Karkaria 2014.

8. "Planning Commission Draft Report: Faster, Sustainable and More Inclusive Growth, An Approach to the Twelfth Five Year Plan (2012–17)," Government of India, Aug. 20, 2011, http://planningcommission.gov.in/plans/planrel/12appdrft/approach_12plan.pdf.

9. See McKinsey Global Institute 2009.

10. See Cushman & Wakefield 2014.

11. As we wrote in Anand and Rademacher (2011), over half of Mumbai's population lives in settlements that occupy only eight percent of the city's area (McFarlane 2008). Eighty percent of these live in homes of less than 100 square feet. (Sanyal and Mukhija 2001). Under the auspices of the Slum Redevelopment Authority and other initiatives, the city faces an enormous and contested rehousing challenge. See also note 5.

12. See, for example, Patankar et al. 2010.

13. Ibid. Although set in a very different regional and political context, Zeiderman 2016 offers an instructive study on the sociopolitical intersection of risk, security, and urban vitality.

14. See, for example, Fuchs 2010.

15. For discussions of expertise in this sense, see Boyer 2008; Carr 2010; Jasanoff 2003; Mitchell 2002.

16. In a now-classic work, Simone (2004) explored the city as a process always inflected with the work of aspiration and imagination. In this case, specific publics were fashioned

in the social experience of collective aspiration, regardless of its ultimate outcome in the material form of the city.

17. The planning and consulting firm Group SCE India Pvt. Ltd. (a French firm with Bangalore-based India offices) was appointed by the MCGM (Municipal Corporation of Greater Mumbai) to develop preparatory studies for the Development Plan (DP). Once the preparatory studies were put into the public domain, several points were raised and considered by the MCGM. Thereafter, a consultant was appointed to prepare the Draft DP 2014—34. This was prepared by EGIS Geoplan Pvt. Ltd. (formerly Group SCE India Pvt Ltd.) in collaboration with an MCGM team, under the leadership of Mr. Vidyadhar Pathak, former Chief of the MMRDA (Mumbai Metropolitan Regional Development Authority). After this DP was published, public outcry underlined several discrepancies within it. A revised committee was then established to revise the DP, headed by a retired MCGM officer.

18. See, for example, the discussion of civic responsibility and sustainability in the editors' introduction in Rademacher and Sivaramakrishnan 2013.

19. See, for example, Rao 2008.

20. Various firms have prepared projections for future new floor space. According to Rawal et al. 2012, for example, in the next eighteen years, India will add 67% of the floor space projected for 2030, or about 2.3 billion square meters.

21. See Rawal et al. 2012.

22. See, for example, McLeod 1983; Hall 2002; Anker 2010.

23. Through this figure, Marx quite famously proposed relations between human life and the non-human natural world. This familiar, oft-quoted passage set the stage for decades of theoretical and political rethinking, and yet revisiting it afresh grounds us in enduring puzzles. The first volume of *Capital* (1967 edition) states "Labour is, in the first place, a process in which both man and Nature participate, and in which man of his own accord starts, regulates, and controls the material re-actions between himself and Nature. . . . By thus acting on the external world and changing it, he at the same time changes his own nature. He develops his slumbering powers and compels them to act in obedience to his sway. . . . We presuppose labour in a form that stamps it as exclusively human. A spider conducts operations that resemble those of a weaver and a bee puts to shame many an architect in the construction of her cells. But what distinguishes the worst architect from the best of bees is this, that the architect raises his structure in imagination before he erects it in reality. At the end of every labour process we get a result that existed in the imagination of the labourer at its commencement. He not only effects a change of form in the material on which he works, but he also realizes a purpose" (pp. 177–78). Quite simply, through labor, Marx considered human beings more as architects than bees. The declaration was powerful in part for its assumption, consistent with existing understandings of bees at the time, that bees did not inhabit sophisticated sensory worlds. See Harvey 2000.

24. Social structures powerfully conditioned, and even confined, consciousness for Marx, shaping the potential of the human imagination to glean a true sense of the limitlessness of the possible. Whether or not full liberation from structurally conditioned modes of thinking can actually be realized remains a core, enduring question around which generations of social analyses continue to organize, albeit with considerable distance from Marx. As Harvey wrote, "we often seem to oscillate in our understanding of ourselves and in our ways of thinking between an unreal fantasy of infinite choice and a cold reality of no

alternative to the business as usual dictated by our material and intellectual circumstances" (Harvey 2000, 204).

25. An exemplary illustration of the importance of temporality in environmental anthropology is assembled in a special issue of the *Journal of the Royal Anthropological Institute* entitled "Environmental Futures" (Barnes 2016). See also McKay 2012 on temporality and humanitarianism, Hetherington 2014 on temporality and development, and Bear 2014.

26. See Appadurai 2013.

27. As I have noted elsewhere (Rademacher 2011), Sivaramakrishnan (1999) points out that for Bourdieu (1990, 52), "practice" is "the site of a dialectic between *opus operandum* and *modus operandi,* of the objectified products and incorporated products of historical practice." My consideration of cultural languages of power as part of structure and practice takes cultural meanings themselves to be "produced in social life and permeated by power relationships that are organized in time and space" (Faure and Siu 1995, 213). The interplay of structure and practice hinges on the analytical concept of the human agent. My own study follows Faure and Siu (1995, 218) in their view of the interaction between culture, history, and agency: "history is created and made significant by meaningful, purposeful actions. However compelled human actions and their unintended results may seem to be, social order and change are not guided by immutable laws. Our view of the human agent as the motive force of history treats culture not as an existing repertoire of values that generations learn and practice, but as a process produced in the flux of social life."

28. As in my previous work, my analytical posture toward social structure and human agency is grounded in the work of Philip Abrams (1982). His idea of "structuring" invites a process-oriented understanding of the "paradox of human agency" (1982, xiii—xiv).

29. See Harvey 2000.

30. See Whatmore 1999.

31. For example, see Candea 2010; Ryan 2012; Singer 2014; Ogden et. al. 2010.

32. See Chakrabarty 2009.

33. See, for example, Ogden et. al. 2010.

34. See, for example, Gandy 2010.

35. I note here the use of these terms as shorthand, but caution that both are complex and connote precise histories and have disciplinary implications.

36. As Keith Murphy has shown in his recent ethnographic work on design as a social process, "design" is conceptually expansive, and, as such, poses a particular kind of analytical problem. He writes, "design itself isn't really a single term, but a collection of homonyms, each of which bears some semantic resemblance to the others, but all of which cover rather different terrain. When we talk about design, we tend to assume we're all really talking about The Same Thing, even if we're not, and this contributes to a fair amount of cross-talk when we collectively think hard about design and its possibilities. I care about this because I wandered to design from other places, and when I landed there, the situation was confusing to me. Design was about things to some people, and practices to others. Or forms and aesthetics. Or systems engineering. Or capitalism. Or collaboration and creativity. Or "what it means to be human." And so on." (Murphy 2016). See also Murphy 2015.

37. Within North American environmental studies, sustainability, particularly in design, was initially widely embraced as a "revolutionary" paradigm shift (Edwards 2005) that promised an entirely new set of technological and cultural norms (McDonough and

Braungart 2002). Architecture, and its green practitioners, were sometimes regarded as part of the potential vanguard of this movement (Gissen 2003; Williamson 2002; Leach 1997; Buchanon 2005).

38. See Rademacher and K. Sivaramakrishnan 2013; Rademacher and K. Sivaramakrishnan 2017.

39. See Rademacher and K. Sivaramakrishnan 2013.

40. See Rademacher and K. Sivaramakrishnan 2013.

41. See Cadenasso and Pickett 2013.

42. Since 1980, the United States National Science Foundation has supported long term ecosystem research at several sites in North America (http://www.lternet.edu/). Two of these are expressly urban sites: the Baltimore Ecosystem Study (http://www.lternet.edu/sites/bes) and Central Arizona-Phoenix Long Term Ecosystem Study (http://caplter.asu.edu/). Both urban LTER sites maintain extensive online libraries of data and analyses.

43. Pickett, Cadenasso, and McGrath 2013.

44. See, for example, Pickett et al. 2001; Rebele 1994.

45. See, for example, the Burch-Machlis Human Ecosystem model as presented in Pickett et al. 1997.

46. See, for example, Turner and Robbins 2008.

47. See Alberti et al. 2003; Collins et al. 2000; Machlis, Force, and Burch 1999; Pickett 1997.

48. See Guy and Moore 2005; Campbell 1996.

49. See Ingersoll 1996.

50. See Rademacher 2015. See also Lawhon et al. 2014.

51. See, for example, Haraway 1989, 1991, 1997; Demeritt 1994; Latour 1993; Swyngedouw 1996; Zimmerer 2000.

52. See, for example, Jasanoff 2004.

53. See Lefebvre 2003.

54. See, for example, Harvey 1973; Brenner 2017.

55. For a rich ethnographic treatment of the ways that ideas and experiences of the "urban" and the "rural" intersect in everyday life, see Harms 2011.

56. Early observations of the untenability of a nature/culture, and by extension nature/city, divide, include now classic work that ranges from pieces such as Cronon's (1995) "The Trouble with Wildnerness" to Latour's (1993) *We Have Never Been Modern*.

57. See, for example, Ong 1999 and Sassen 1991.

58. See, for example, Kaika 2005; Swyngedouw 1996, 1999; Gandy 2002; Kabirh 1984; Castells 1996.

59. See Baviskar 2003.

60. See, for example, Mitchell 2002; Tsing 2000, 2012.

61. See, for example: Castree and Braun 2001; Braun 2005.

62. Science and Technology Studies (STS) has paid productive attention to the social dynamics of scientific knowledge production (Dumit 2004; Downey and Dumit 1997; Franklin 1997; Hogle 1995; Stengers 1993; Rabinow 1992), rigorously demonstrating how situated actions and contingent decisions characterize scientific work (e.g., Knorr-Cetina 1981; Latour & Woolgar 1986); this book extends this line of inquiry to architectural practice and problem solving. Work in STS has also shown that technical problems are often defined in

relationship to spaces in which, and processes through which, specific forms of knowledge are produced (Callon 1995). This book aims to contribute to work in STS that explores disciplined ways of organizing and making sense of the natural world (Barnes, Bloor, and Henry 1996; Gooding 1992; Lynch 1985) by asking, how do architects acquire, and organize, the "ecological" knowledge that forms the basis of their practices? What "counts" as ecology?

63. See for an overview Orr, Lancing, and Dove 2015; Franklin 1995; Fischer 2007; Latour 1988; Callon 2009.

64. See Choy 2011.

65. See Chakrabarty 2009.

66. See Rademacher and Sivaramakrishnan 2013.

67. For instance, an extensive literature illustrates the complex ways that postcolonial modernities and cultures shape the built environment. King 2004, and authors like Hosagrahar (2005), Chattopadhyay (2012), and Rajagopalan and Desai (2012) refine, critique, and enrich debates about the ways that architecture and urbanism intersect with colonial, nationalist, and modernizing projects.

68. It is helpful here to invoke Foucault's now-classic discussion of architects, architecture, and space in *Space, Knowledge, Power*. Here, we are reminded to think of the built environment as in many ways a product and mechanism of Foucault's idea of power-knowledge. The agency of individual architects is embedded in a web of knowledge forms, social structures, and cultural norms that constitute power-knowledge relations and histories. Often, the potential for any form of meaningful agency is fully negated by those same relations.

69. Anthropologists have long taken interest in the imagined relationship between the built form and social form, as well as social change, social harmony, and social process. An early review of these efforts may be found in Lawrence and Low (1990). Buchli (1999) adds important insights to these engagements by showing "how seemingly weighty, inscribed, and totalizing world views (Blier 1987) or 'spatial logics' (Hillier and Hanson 1984) can be radically subverted." Buchli's work notes that anthropologists, most often those working in studies of material culture, have formally and carefully recognized a tendency to "posit a direct, iconic, and at times homologous correspondence between an item of material culture and the society with which it is associated," and that we must attend to the ways control can be exercised at multiple social scales (Foucault 1977; Shanks and Tilley 1987, 1992). Buchli's work directly addresses the enduring analytical dilemma of identifying the parameters of human agency, describing the challenge "to overcome the image of (human actors) enslaved to the fixed meanings and deterministic structures of a given society, where individuals were seen to respond in a mechanistic, and ultimately helpless, fashion to irresistible structural prerogatives as in the unilineal and deterministic tradition in Morgan, Marx and Engels, and all the way through to structuralism (Levi Strauss 1966; Chomsky 1968; Glassie 1975; Deetz 1977)" (Buchli 1999, 8). By layering ecosystems and ecosystem processes into the analytical calculus, we face the complicated juxtaposition of systems that are understood as relatively mechanistic (that is, biogeochemical systems) and those in social life that, although structurally conditioned, are never automatically pre-configured. See Buchli 1999. See also Fennell 2015.

70. See examples in Hall 2002 and Anker 2010. For an excellent treatment of a contemporary experiment in urban environmental design see Günel 2016 and forthcoming (2018).

71. Here, I use "hybridity" in the sense first proposed in work by Swyngedouw (1996, 2006).

72. See work by Jasanoff and Martello 2004; Mitchell 2002; Bryant 1998; Blaikie 1999; Bocking 2004; Buuren & Edelenbos 2004; Collingridge and Reeve 1986; Davis and Wagner 2003; Dimitrov 2003.

73. See Taylor and Buttel 1992.

74. In *Reigning the River* (2011, 15), I suggested that the making of nature and the simultaneous making of meaningful life in the city involves a complex social identity construction process that may beget "new affinities . . . environmental affinities that might 'foster' cohesion where other ways of marking sameness and difference (cannot)." These affinities may reveal the many dimensions of existing identity struggles, contests over governance, and collective reworkings of the moral ecologies of city living. In this sense, the 'places' of nature in the city are always in a state of refashioning (see also Rademacher and Sivaramakrishnan 2017).

75. See Braun 2005.

76. See Appadurai 2013.

77. Jyoti Hosagrahar (2005, 55) explores anxieties about identity in Indian architecture, observing, "the revised canon about Indian architectural history continues to be steeped in a 'message of ancient and medieval greatness' and highlights reasons like the role of institutional bureaucracy." Her analytically important references to the control of architectural education by such bodies as the All India Council for Technical Education, the AICTE, remain relevant to an always-changing institutional landscape. In 2009, for example, AICTE was reformed with important implications.

78. Paniker (2008, 58–62) makes a useful distinction between a "discourse level" (architectural historiography) and an "institutional level" (architectural education) in the production of architectural knowledge in India.

79. Building Research Establishment Environmental Assessment Method (BREEAM).

80. Leadership in Energy and Environmental Design (LEED).

81. Green Rating for Integrated Habitat Assessment (GRIHA).

2. THE INTEGRATED SUBJECT

1. See Cadenasso and Pickett 2013.

2. Rachana Sansad "Vision" page, http://rachanasansad.edu.in/.

3. At the time of the work, the total fees per student per year were Rs. 87,000 (which included a refundable library deposit of Rs. 4000). RSIEA provides concessions for students whose financial circumstances prevent them from paying in full immediately. A few scholarships were available for students in their second year, but these did not constitute the bulk of tuition.

4. Roshni Udyavar Yehuda, personal communication, March 2012.

5. A portion of the RSIEA website reads, "The Research and Design Cell of the Institute provides independent consultancy wherein a team of students guided by faculty members undertakes projects. Some past projects include Ecotourism projects in Sawantwadi, Restoration of Charolette lake at Matheran, Dahisar River Restoration Project, Environment Improvement Projects in slums such as Behrampada and Mahatma Phule Nagar, Mumbai, &

Rainwater harvesting for several housing societies and corporate houses including HPCL, Mumbai office." See http://rachanasansad.edu.in/.

6. Dr. Ashok Joshi, interview transcript, March 2012.

7. Ibid.

8. Ibid.

9. See http://rachanasansad.edu.in/.

10. Dr. Ashok Joshi, interview transcript, March 2012.

11. This follows work by Goldman (2001) and Li (2008) on eco-rational subjects, as well as Agrawal's (2005) application of these ideas directly to an environmental domain, demonstrating the utility of the concept of "environmentality" as a social process that is constitutive of environmental subjectivity. A range of work has followed in this vein, as environmental anthropologists have noted the ways that certain configurations of institutions, knowledge, and politics relate in turn to subjectivities that reinforce certain conceptual categories and notions of proper care for the natural environment. An excellent recent account may be found in Mathews (2011).

12. Field notes, July 12, 2012.

13. Ibid.

14. Ibid.

15. The literature on commensurability in political ecology studies is vast, but an instructive starting point may be found in the work of J. Martinez-Alier, particularly 2004.

16. Field notes, July 2012.

17. Roshni Udyavar-Yehuda, Rajeev Taischete, and Mukund Porecha were also partners in Enviro-Arch, an environmental architecture firm.

18. Field notes, April 17, 2012.

19. Field notes, April 8, 2012.

20. Text of email message dated October 10, 2014.

21. Udyavar and Shah 2010.

3. ECOLOGY IN PRACTICE

1. Text of PowerPoint slide presented by Dr. Doddaswmy Ravishankar, Opening Day Ceremony, RSIEA, 2012.

2. International Society for Krishna Consciousness (ISKON).

3. See Paniker 2008, 113.

4. Dr. Ashok Joshi, interview transcript, March 2012.

5. See Carson 1962; Lovelock 1979; Hernandez and Mayur 2010.

6. Through his Gaia Hypothesis, James Lovelock (1979) advanced the idea that the Earth's biosphere is usefully conceptualized as an organism, the constituent parts of which constitute mechanisms for self-regulation. RSIEA course material introduced this idea without endorsing it as fact or fiction. It was invoked to mark an influential way of applying systems ecological thinking to the scale of the Earth's biogeochemical complex.

7. Jasanoff 2004a, 2004b; Latour 1993; Mathews 2011.

8. Mathews (2011, 23) notes that a rich literature in science and technology studies outlines these points more substantively. He notes in particular how Gieryn (1995) argued that the boundary between political and technical domains is constantly reworked and contested, while Hilgartner (2000) argued that scientific advice is always in some way shaped

by the assessments of its audience. Any application of scientific knowledge in practice may be seen, in this way, as automatically public—a performance of expertise.

9. An instructive starting point for thinking about such forms of interdisciplinary borrowing may be found in Dove 2001.

10. Field notes, December 11, 2008, during Green Home Technologies Lecture, Auroville.

11. Ibid.

12. Ibid.

13. Carrying capacity in this sense is generally defined as the total population of living organisms that a defined habitat area can support.

14. RSIEA "New Curriculum" Program document.

15. Quotations in this section are derived from field notes, March 17, 2010.

16. Field notes, March 17, 2010, in a class meeting of "Sustainable Building Design Principles."

17. See http://www.unep.org/ietc/SATAssessing/tabid/56441/Default.aspx.

18. For an instructive critical account of indicators and metric formulas in arenas of governance, see Merry, Davis, and Kingsbury 2015.

19. See Elkington 2001.

20. See http://www.environdec.com/en/What-is-an-EPD/#.VFJNrNwtCBg.

21. Over the past decade, standard Euro-American metrics for assessing the sustainability of built forms, such as LEED and BREEAM, have been engaged, contested, sometimes reinforced, and sometimes reworked, in India. The U.S. Green Building Council, which developed LEED standards, played a foundational role in establishing the World Green Building Council, a consortium that includes national councils worldwide. Among these is the Indian Green Building Council (IGBC). Concerns among Indian architects and builders that these standards were not always the most appropriate measures for built form sustainability, as well as reservations about the relationship between the IGBC and the Indian construction industry, led to the development of alternative national and regional guidelines for green design; examples include ECOHOUSING and GRIHA. Localized alternative metrics such as these, and the ways they are used to challenge the IGBC, exemplify the many ongoing contests between "local" and "extralocal" ideas and practices of green architecture.

22. Quotations from field notes, December 11, 2008, during Green Home Technologies Lecture, Auroville.

23. Quotation recorded in field notes, December 8, 2008, Grundfos Manufacturing Chennai, Regional CEO.

24. Ibid.

25. Field Notes, Design Studio course, March 18, 2010.

4. RECTIFYING FAILURE

1. Readers with a particular interest in the complex politics and multi-scaled audiences that spectacles such as the ones described in this chapter attempt to reach may find these topics more fully addressed in the chapter, "Emergency Ecology and the Order of Renewal," in Rademacher 2011.

2. See Anand and Rademacher 2011 for a fuller discussion.

3. Many figures exist. This figure was published by a Special Commission appointed by the Indian Housing and Urban Poverty Alleviation Ministry, and reported in the Times of India. See "City of Dreams?" Singh, Mahendra Kumar, *Times of India*, November 15, 2010.

4. See Anand and Rademacher 2011.

5. United Nations. 2011 Revision, World Urbanization Prospects.

6. Government of India Census, 2011. Readers should note that the overall reliability of such statistics has been called into question. For example, see Agrawal and Kumar (2014).

7. See Anand 2017.

8. In a twenty-four hour period, 994 mm, or 39.1 inches of rain fell on Greater Mumbai.

9. See, for example, Baviskar 2011.

10. See, for example, D'Souza 2002, 2006; D'Souza, Mukhopadhyay, and Kothari 1998.

11. See Daud 2011, 207.

12. See also Villiers-Stuart 1913.

13. On Victorian gardens, see Morgan and Richards 1990.

14. The firm's website may be reviewed here http://www.pkdas.com/.

15. Mumbai Waterfronts Center and PK Das & Associates 2012.

16. Ibid, 1.

17. Ibid, 3.

18. Ibid, 5.

19. Ibid, 7.

20. Seafront promenade projects at Bandra Bandstand and Carter Road, as well as the Gateway of India, were highlighted with a dramatic before-and-after photo; Juhu Beach illustrated what was possible, while Dadar-Prabha-Devi Beach exemplified a dire beach conservation and nourishment problem.

21. Mumbai Waterfronts Center and P.K. Das and Associates. 2012, 34.

22. See, for example, Baviskar 2003b, 2009; Rademacher 2009; Ghertner 2013; Doshi 2013; Sharan 2014.

23. According to the group's website, "CitiSpace (Citizens' Forum for Protection of Public Spaces), established in June 1998, is an NGO which networks over 600 Resident Associations, Community Based Organisations (CBOs), NGOs, Trade/ Commercial Establishments and individuals in most of Mumbai's 24 Wards. Our creed is the protection of all Public Open Spaces (such as Footpaths, Playgrounds, Recreation grounds, No Development Zones, Beaches and Mangroves, etc.) and advocacy of the rightful use of those spaces." (http://nagaralliance.org/citispace/).

24. This NGO had over fifteen years of advocacy experience among slum dwellers and open space advocates, highlighting the political work of open space provision in contrast to the *Open Mumbai* emphasis on an aspirational imaginary.

25. The survey was led by Neera Punj and Nayana Kathpalia of CitiSpace, and assisted by architects and architecture students. The website reads, "In 2008 CitiSpace undertook a survey of Reserved Public Open Spaces which was completed in 2013 with about 1800 spaces surveyed. The first phase of the Survey was published in the book entitled *Breathing Space: A Fact File of 600 Reserved Public Open Spaces of Greater Mumbai in June 2010.*"

26. Accessed at http://nagaralliance.org/citispace/2012.

27. See Rao 2013, 158.

28. Ecology is usually traced to its etymological origins in the Greek *oikos*, or home, and *–ology*, or "the study of."

29. The article continued: "A desperate fire department threw in everything they had in battling the blaze, pressing 26 fire engines into service and managed to evacuate nearly 3,000 employees but could not prevent the blaze from completely destroying the top three floors of the state government's main administrative building."

30. The Adarsh Housing Society scandal was publically exposed in November 2011, when a report by the Comptroller and Auditor General of India detailed how various elites from political, bureaucratic, and military domains conspired to alter urban development and construction regulations in the course of building the Adarsh Housing Society in Colaba. In the process, they ensured for themselves luxury flats at rates well below-market value. In the scandal's wake, then-Chief Minister of Maharashtra, Ashok Chavan, resigned his post.

31. See Rao 2013, 155.

32. Ibid, 27.

33. Ibid, 54.

34. See Rao 2013,145 and the section entitled, "Architects and the Profession" for a more detailed account of factors such as the rise and consolidation of an Indian middle class, the arrival of reinforced concrete cement technology, and important design modifications like the introduction of the toilet inside the flat.

35. "Even though the Trust required that all builders get their plans approved by an architect, the architect's role was usually quite superficial at the time." (Rao 2013, 145).

36. See Rao 2013, 145–46.

37. This figure is according to the 2001 Government of India Census.

38. See Rao 2013, 204–5 for a list of specific responsibilities.

39. Laxmi Deshmukh, interview transcript, *All India Institute of Local Self Government*, March 26, 2012.

40. Ibid.

41. Ibid.

42. Ibid.

43. Accessed at http://www.ft.com/cms/s/0/548a4e60-c11b-11da-9419-0000779e2340.html#axzz3EzC28MLD.

44. Laxmi Deshmukh, interview transcript, *All India Institute of Local Self Government*, March 26, 2012.

45. Laxmi Deshmukh, interview transcript, *All India Institute of Local Self Government*, June 25, 2012.

46. Ibid.

47. Laxmi Deshmukh, interview transcript, *All India Institute of Local Self Government*, March 26, 2012.

48. Laxmi Deshmukh, interview transcript, *All India Institute of Local Self Government*, June 25, 2012.

49. Text to come

5. MORE THAN HUMAN NATURE AND THE OPEN SPACE PREDICAMENT

1. See, for example, Economist Intelligence Unit, 2011, "Asian Green City Index: Assessing the Performance of Asia's Major Cities." Munich: Siemens. Accessed at http://www.siemens.com/entry/cc/en/greencityindex.htm.

2. Ibid.

3. See, for example, Baviskar 2003b, 2009; Rademacher 2009; Ghertner 2013; Doshi 2013.

4. See, for instance, Rademacher and Sivaramakrishnan 2013.

5. The City of New York. 2013. *PlaNYC Progress Report 2013*, 16.

6. See, for example, Baviskar 2003b, 2009; Rademacher 2009; Ghertner 2013; Doshi 2013.

7. The Parsis are a Zoroastrian minority group concentrated in Gujarat and Sindh. In Mumbai, the group has roughly sixty thousand members, and a general decline in the overall population characterizes the past several decades.

8. See Bombay Natural Historical Society 2012, 4.

9. "Lose the vultures, lose the soul." Karkaria, B. *New York Times,* May 11, 2007.

10. It should be acknowledged that embedded in this moment of imagining a future Mumbai, the city's Parsi community is famous for its own anxiety about its future. Although a thorough treatment of this issue is beyond the scope of this chapter, the fact that many regard Parsi religious and cultural identity as itself "endangered" gave a particular valence to the future of the Towers of Silence and the Doongerwadi forest. See Axelrod 1990.

11. Personal communication, March 2012.

12. A vast literature captures a range of issues related to ecological concerns and debates about non-native species. Some instructive articles include Davis and Thompson 2000; Gurevitch and Padilla 2004; Lodge 1993; Mack et al. 2000.

13. A classic starting point for understanding this concept is Costanza 1997. See also Boyd and Banzhaf 2007; DeGroot et al. 2010; and Farber, et al. 2006. More recent critical treatment of ecosystem services as a concept and bundle of practices abounds; an excellent entry point to this literature is Ernston 2013.

14. A classic starting point for understanding the concept of disturbance in ecosystem ecology is Pickett and White 1985.

15. Here, Harvey's (1999) "Considerations on the Environment of Justice" is instructive. In pointing to the inherent contradictions of the idea of a universal environmental ethnic, he argues that it is simultaneously impossible, desirable, and inevitable. The inevitability is conditional, however, and fully reliant on our human social capacity for what he calls a more honest mode of translation—one in which the terms and lifeways that frame another person's experience of the environment is a starting point for communication and analysis. The advocate of environmental justice, he argues, must be constantly self-reflexive and humble.

16. Following the concept of *ecologies of urbanism* developed in partnership with K. Sivaramakrishnan (2013, 2017), I mean to signal here that there may be many ways of valuing, and naming, the multiplicity of forms of life on Earth. These may go unrecognized for their overlap with the content of the natural scientific and policy term, "biodiversity." The more expansive content of the concept may be captured in other ways of knowing nature, and its value may be differently designated in those multiple knowledge forms.

17. See "Ways of Knowing Nature" in Rademacher 2011.

18. Heynen 2003, as quoted in Braun 2005, 645.

6. CONSCIOUSNESS AND INDIAN-NESS

1. Pondicherry, December 14th lecture to RSIEA students.

2. Ibid.

3. In *Reigning the River* (2011, 15), I suggested that the making of nature and the simultaneous making of meaningful life in the city involves a complex social identity construction process that may beget "new affinities . . . environmental affinities that might 'foster' cohesion where other ways of marking sameness and difference (cannot)." These affinities may reveal the many dimensions of existing identity struggles, contests over governance, and collective reworkings of the moral ecologies of city living. In this sense, the "places" of nature in the city are always in a state of refashioning (see also Rademacher and Sivaramakrishnan 2017).

4. See for general grounding on this issue Ghassam-Fachandi 2012; Mishra 2006; Davis 2005; Jaffrelot 1993, 1998; Hansen 1999; Basu et al. 1993; Bhatt 2001; Bhagavan 2010.

5. Paniker (2008, 82) also notes that in 1984, the Indian National Trust for Architectural and Cultural Heritage was established, identifying its main objective as the restoration and conservation of "neglected" art and cultural heritage in India; Paniker interprets this as "the making of public meaning and belief about the Indianness of things."

6. Paniker 2008, 74.

7. Anonymized student reflection on returning from Auroville, transcript.

8. Promotional pamphlet introducing *Govardhan Eco-village.*

9. Invoking cow protection, though treated as largely politically benign in the context of this field study visit, is laden with political symbolism, much of it associated with various strains of Hindu Nationalism. See for example Ghassem-Fachandi 2012; Hansen 1999; Pandey 1983; Freitag 1980.

10. Anonymized student reflection on returning from Auroville, transcript.

11. Anonymized student reflection on returning from Auroville, transcript

12. Rao 2013, 146.

13. Ibid, 146–7.

14. Ibid, 147.

15. Ibid, 148.

16. Rao 2013, 150.

17. Sears 2001, 133.

18. Ibid, 134.

19. Ibid.

20. Ibid, 136.

21. Ibid.

22. Field notes, Auroville, December 10, 2007.

23. Auroville's own public relations material, as reported on its website, lists a Governing Board, a Residents Assembly (comprised of the current full members of the Auroville community), a Working Committee, an International Advisory Council, and a Secretary of the Auroville Foundation. The International Advisory Council was established by the Government of India; it appoints the Council members. In the past, the Council has included such notable figures as Amartya Sen. See http://www.auroville.org/.

24. In his *From Bauhaus to Ecohouse,* Anker (2010: 167–8) notes that McHarg's work is sharply critical of Western capitalist individualism and greed. As a remedy, McHarg proposes an Orientalist, holistic ecology as its remedy (Anker 2010, 167–68).

25. Anonymized student reflection on returning from Auroville, transcript.

26. Anonymized student reflection on returning from Auroville, transcript.

27. Anonymized student reflection on returning from Auroville, transcript.

28. Anonymized student reflection on returning from Auroville, transcript.

29. Sears 2001, 137.

30. Accessed at http://www.ecovillage.org.in/perspectives/land-and-cow-a-perfect-sustainable-system/.

31. A detailed description of the system, excerpted from the Govardhan eco-village website: "SBT system consists of an impervious containment and incorporates soil, formulated granular filter media, select culture of macro organisms such as earthworms and plants. It involves a combination of physical and biological processes for processing of wastewater and it derives its fundamental principle from the functioning of a terrestrial ecosystem. The process by design integrates with the natural bio-geochemical cycles of nature and hence proves to be most effective. The combined grey and black water from all the residential facilities are collected and transported via a water based underground sewerage network to a central collection point. In the first stage the physical separation of waste is accomplished in a primary treatment unit consisting of a perforated screen and gravity-settling tank and an equalization tank. The perforated screen helps in separating the undissolved solid wastes from the waste water and allows it to pass through a settling chamber that has a sloped bottom opposite to the direction of the water flow, thus facilitating the settling of solid particulates with higher specific gravity than the waste stream. Then the water enters the open top equalization tank that allows the dissolved pollutants to be exposed to natural sterilization by sunlight and ambient air. In this second stage the wastewater is sprayed, by means of a pump, onto a plant bed which is part of an engineered ecosystem that constitutes two bioreactors, one for a coarse purification and the other for further refining through recycling. This ecosystem consisting of soil, bacterial culture and earthworms, mineral additives and select plants, treats the water in a combination of physio-chemical and biological processes. Purification takes place by adsorption, filtration and biological reaction. The entire waste is processed and converted into bio-fertilizer which is rich in organic content, and is being used in the plant nursery at GEV. The other useful by-product is the Biomass in the form of flower, fodder, fruit and fiber which are also completely utilized in house. Since the entire waste is converted, there are no issues like handling the wastes after treating the water, as is common in conventional chemical based sewage treatment plants. The entire process operates in aerobic mode thus eliminating the possibility of foul odor near the plant, creating a safe and serene ambiance for the people dwelling near the plant. The processed water can be reused in gardening, agriculture and also supports marine life. The SBT plant at GEV can handle up to 30,000 litres of sewage per day and operates in an 8 hour cycle daily. It can potentially produce up to 20,00,000 Tons of bio-fertilizer per year and most importantly offers an eco-friendly option to the growing menace of waste handling." (See: http://www.ecovillage.org.in/perspectives/a-flush-story-iii-soil-biotechnology-plant-at-govardhan-eco-village/).

32. Examples include Desai and Rajagopalan 2012; Rajagopalan 2012; Rajagopalan 2011.

33. Sears 2001, 136. The question of whether the use of vernacular forms can be inclusive rather than exclusive is also posed by Hasan 2001.

34. Sears 2001, 137, quoting Meister (9–15) in the same volume.

35. *Architectural Review*, August 1987.

36. Cruickshank's Introduction to the special issue of *Architectural Review* is cited by Paniker (2008, 97).

37. Curtis 1987, as cited by Paniker 2008, 98.
38. Curtis 1988, as cited by Paniker 2008, 102.

7. A VOCATION IN WAITING

1. Anonymized quotation from a current (2012) RSIEA student survey response.
2. Field notes from group interview at RSIEA, June 24, 2012.
3. Ibid.
4. Anonymized quotation from a current (2012) RSIEA student survey response.
5. Siddharth currently works in an architecture firm that does not undertake specifically environmental architectural projects. It employs 70–80 architects, and serves clients all over India. At the time of this interview Siddharth was doing a residential development project in Kochi, on reclaimed land. Siddharth told me that the client wants a gold certification for this development.
6. Siddharth, interview transcript, March 2012.
7. Aditya, interview transcript, April 2012.
8. Ibid.
9. Darius, interview transcript, February 2012.
10. Ibid.
11. Ibid.
12. After graduation, Amrit worked for two years with what he called "typical commercial architects," but then left the firm to seek environmental opportunities; at the time of our interview, he was in private practice, busy primarily with designing bungalows in Ali Baug.
13. Amrit, interview transcript, April 2012.
14. Ibid.
15. Anonymized quotation from a recent (2009) RSIEA graduate survey response.
16. Siddharth, interview transcript, March 2012.
17. Darius, interview transcript, February 2012.
18. Aditya, interview transcript, April 2012.
19. Darius, interview transcript, February 2012.
20. Aditya, interview transcript, April 2012.
21. Kalpana, interview transcript, April 2012.
22. Field notes from group interview at RSIEA, June 24, 2012.
23. Siddharth, interview transcript, March 2012.
24. Suhasini, interview transcript, March 2012.
25. Field notes from group interview at RSIEA, June 24, 2012.
26. Aditya, interview transcript, April 2012.

8. SOLDIERING SUSTAINABILITY

1. I am grateful to K. Sivaramakrishnan for encouraging me to consider this point.
2. The work of Guha and Martinez-Alier (1997) provides a useful overview here, but there are vast literatures on all of these points. Again, I gratefully acknowledge K. Sivaramakrishnan for highlighting this point.

3. See, for instance, Srinivas 2015.

4. For example, see Pickett, Cadenasso, and McGrath 2013.

5. To further note that many good design examples were derived from non-city settings highlights the recurrence of a rather stark experiential binary in which nature in its most intact forms is sought in places separate from the city rather than embedded within them. Environmental architects in training went to nature—a "nature" removed from the city, and in many ways the city's opposite—in order to learn environmental architecture. They actively sought guidance for the city of the future by leaving it altogether, in search of purer forms of nature.

6. Appadurai 2015, 481.

7. See Appadurai 2000. As Appadurai (2015, 482–3) further notes, this speculative territory is the "zone where the visible and the invisible come together," the joining of the visible city to the processes that will activate its now invisible future form.

8. See Buck-Morss 2000; Humphrey 2005 for a more detailed treatment of the relationship between utopian aspiration, built form, and ideology.

9. I draw here from Sivaramakrishnan (2015) and Pandian and Ali's (2010) *Ethical Life in South Asia,* which emphasize the moral dispositions and everyday lived practices that characterize social life.

10. See, for example, Connolly 2013; Barua 2014; Whatmore 2003.

REFERENCES

Abrams, P. 1982. *Historical Sociology*. New York: Cornell University Press.

Agarwal, A., and S. Narain, eds. 1997. *Dying Wisdom: Rise, Fall, and Potential of India's Traditional Water Harvesting Systems*. Delhi: Center for Science and Environment.

Agrawal, A. 2005. *Environmentality: Technologies of Government and the Making of Subjects*. Durham: Duke University Press.

Agrawal, A., and V. Kumar. 2012. "How Reliable are India's Official Statistics?" East Asia Forum, April 6.

Alley, K. 2002. *On the Banks of Ganga: When Wastewater Meets a Sacred River*. Ann Arbor: University of Michigan Press.

Alberti, M. et al. 2003. Integrating Humans into Ecology: Opportunities and Challenges for Studying Urban Ecosystems. *BioScience* 53: 1169–70.

Ali, D. 2011. Gardens in Early Indian Court Life. In *India's Environmental History: From Ancient Times to the Colonial Period*, ed. M. Rangarajan and K. Sivaramakrishnan, 182–214. New Delhi: Permanent Black.

Anand, N. 2017. *Hydraulic City: Water and the Infrastructure of Citizenship in Mumbai*. Durham: Duke University Press.

Anand, N., and A. Rademacher. 2011. Housing in the Urban Age: Inequality and Aspiration in Mumbai. *Antipode* 43: 1748–72.

Anderson, B. 1983. *Imagined Communities*. New York: Verso.

Anker, P. 2010. *From Bauhaus to Ecohouse: A History of Ecological Design*. Baton Rouge: LSU Press.

Appadurai, A. 2015. Afterward: the Dreamwork of Capitalism. *Comparative Studies of South Asia, Africa, and the Middle East* 35 (3): 481.

_____. 2013. *The Future as Cultural Fact*. London: Verso.

_____. 2000. Spectral Housing and Urban Cleansing: Notes on a Millennial Mumbai. *Public Culture* 12 (3): 627–51.

Axelrod, P. 1990. Cultural and Historical Factors in the Population Decline of the Parsis of India. *Population Studies* 44: 401–19.

Barnes, J., ed. 2016. Environmental Futures. Special Issue of *Journal of the Royal Anthropological Institute* 22 (S1).

Barnes, B., D. Bloor, and J. Henry. 1996. *Scientific Knowledge: a Sociological Analysis.* Chicago: University of Chicago Press.

Barua, M. 2014. Volatile Ecologies: Toward a Material Politics of Human-Animal Relations. *Environment and Planning A* 46: 1462–78.

Basu, T. et al. 1993. *Khaki Shorts, Saffron Flags.* London: Orient Longman.

Baviskar, A. 2003. For a Cultural Politics of Natural Resources. *Economic and Political Weekly* 38 (2003a): 5051–55.

_____. 2003. Between Violence and Desire: Space, Power, and Identity in the Making of Metropolitan Delhi. *International Social Science Journal* 55 (1).

_____. 2005. Toxic Citizenship: The Quest for a Clean and Green Delhi. Unpublished manuscript.

_____. 2011. Cows, Cars and Cycle Rickshaws: Bourgeois Environmentalists and the Battle for Delhi's Streets. In *Elite and Everyman: The Cultural Politics of the Indian Middle Classes,* 391–418. New York: Routledge.

Bear, L. 2014. Doubt, Conflict, Mediation: The Anthropology of Modern Time. *Journal of the Royal Anthropological Institute* 20 (S1): 18–20.

Bhagavan, M. 2010. The Hindutva Underground: Hindu Nationalism and the Indian National Congress in Late Colonial and Early Postcolonial India. In *Religion and Identity in South Asia and Beyond: Essays in Honor of Patrick Olivelle,* ed. Steven Lindquist. New York: Anthem.

Bhatt, C. 2001. *Hindu Nationalism: Origins, Ideologies, and Modern Myths.* New York: Bloomsbury.

Biswas, A.K., and S.B.C. Agrawal. 1992. *Environmental Impact Assessment for Developing Countries.* London: Butterworth-Heinemann.

Blier, S. P. 1987. *The Anatomy of Architecture.* Chicago: University of Chicago Press.

Bombay Natural Historical Society. 2012. *Setting up a conservation breeding program for two resident Gyps species of vultures, white backed vulture* (Gyps bengalensis) *and long billed vulture* (Gyps Indicus) *at Maharashtra.* Project proposal.

Bourdieu, P. 1990. *The Logic of Practice.* Stanford: Stanford University Press.

Boyd, J., and S. Banzhaf. 2007. What are Ecosystem Services? The Need for Standardized Environmental Accounting Units. *Ecological Economics* 63: 616–26.

Boyer, D. 2008. Thinking through the Anthropology of Experts. *Anthropology in Action* 15 (2): 38–46.

Braun, B. 2005. Environmental Issues: Writing a more than human urban geography *Progress in Human Geography* 29 (5): 635–50.

Brenner, N. 2017. *Critique of Urbanization: Selected Essays.* Basil: Bau Verlag.

Buchanan, P. 2005. *Ten Shades of Green: Architecture and the Natural World.* New York: The Architectural League of New York.

Buchli, V. 1999. *An Archaeology of Socialism.* New York: Berg.

Buck-Morss, S. 2000. *Dreamworld and Catastrophe.* Cambridge: MIT Press.

Burdett, R., ed. 2007. *Urban India: Understanding the Maximum City.* Cities Programme, London School of Economics and Political Science and Alfred Herrhausen Society. Report.

Cadenasso, M. L., and S.T.A. Pickett. 2013. Three Tides: The Development and State of the Art of Urban Ecological Science. In *Resilience in Ecology and Urban Design: Linking Theory and Practice for Sustainable Cities,* ed. S. T. A. Pickett, M. L. Cadenasso, and B. McGrath. New York: Springer.

Caldeira, T. 2001. *City of Walls: Crime, Segregation, and Citizenship in São Paulo.* Berkeley: University of California Press.

Callon, M. 1995. Four Models for the Dynamics of Science. In *Handbook of Science and Technology,* ed. S. Jasanoff, G.E. Markle, J.C. Petersen, and T. Pinch. New York: Sage.

_____. 2009. Some Elements of a Sociology of Translation: Domestication of the scallops and the fishermen of St. Brieuc Bay. In *Power, Action, Belief: A new sociology of knowledge?,* ed. John Law. New York: Routledge.

Candea, M. 2010. I Fell in Love with Carlos Meerkat: Engagement and Detachment in Human-Animal Relations. *American Ethnologist* 37 (2): 241–258.

Canter, L. 1995. *Environmental Impact Assessment.* New York: McGraw-Hill.

Carr, E. S. 2010. Enactments of Expertise. *Annual Review of Anthropology* 39: 17–32.

Carson, R. 1962. *Silent Spring.* New York: Houghton Mifflin.

Castree, N., and B. Braun, eds. 2001. *Social Nature: Theory, Practice, and Politics.* New York: Wiley-Blackwell.

Chakrabarty, D. 2009. The Climate of History: Four Theses. *Critical Inquiry* 35: 197–222.

Chalana, M. 2010. Slumdogs vs. Millionaires: Balancing Urban Informality and Global Modernity in Mumbai, India. *Journal of Architectural Education* 63 (2): 25–37.

Chatterji, R., and D. Mehta. 2007. *Living with Violence: An Anthropology of Events and Everyday Life.* New York: Routledge.

Chattopadhyay, S. 2012. *Unlearning the City: Infrastructure in a New Optical Field.* Minneapolis: University of Minnesota Press.

Chomsky, N. 1968. *Language and Mind.* New York: Harcourt, Brace, and World.

Chopra, P. 2011. *A Joint Enterprise: Indian Elites and the Making of British Bombay.* Minneapolis: University of Minnesota Press.

_____. 2012. Free to move, forced to flee: the formation and dissolution of suburbs in colonial Bombay. *Urban History* 39 (1): 83–107.

_____. 2007. Refiguring the Colonial City: Recovering the Role of Local Inhabitants in the Construction of Colonial Bombay. *Buildings and Landscapes* 14: 109–125.

Choy, T. 2011. *Ecologies of Comparison: An Ethnography of Endangerment in Hong Kong.* Durham: Duke University Press.

City of New York. 2013. *PlaNYC Progress Report 2013.* Report.

Cohelo, K., and N.V. Raman. 2013. From the Frying Pan to the Floodplain: Negotiating Land, Water, and Fire in Chennai's Development. In *Ecologies of Urbanism in India: Metropolitan Civility and Sustainability,* ed. A. Rademacher and K. Sivaramakrishnan, 145–68. Hong Kong: Hong Kong University Press.

Connolly, W. 2013. New Materialism and the Fragility of Things. *Journal of International Studies* 41 (3): 399–412.

Costanza, R. et al. 1997. The Value of the World's Ecosystem Services and Natural Capital. *Nature* 387: 253–60.

Cruickshank, D. 1987. Introduction to the Special Issue. *Architectural Review* 13: 8.

Cronon, W. 1995. The Trouble with Wilderness, or, Getting Back to the Wrong Nature. In *Uncommon Ground*. New York: W.W. Norton.

Curtis, W. J. R. 1987. Modernism and the Search for Indian Identity. *Architectural Review* 1086: 33–38.

_____. 1988. *Balkrishna Doshi: An Architecture for India*. New York: Rizzoli.

Cushman & Wakefield. 2014. Office Space Across the World. Report, Cushman & Wakefield Corporate Communications.

Davis, M. A., and K. Thompson. 2000. Eight Ways to be a Colonizer; Two Ways to be an Invader: A Proposed Nomenclature Scheme for Invasion Biology. *Bulletin of the Ecological Society of America* 81 (3): 226–30.

Davis, M. 2006. *Planet of Slums*. New York: Verso.

Davis, R. H. 2005. The Cultural Background of Hindutva. In *India Briefing: Takeoff at Last?*, ed. A. Ayres and P. Oldenburg, 107–40. New York: Asia Society.

Deetz, J. 1977. *In Small Things Forgotten*. New York: Anchor.

DeGroot, R.S. et al. 2010. Challenges in Integrating the Concept of Ecosystem Services and Values in Landscape Planning, Management, and Decision-making. *Ecological Complexity* 7: 260–72.

Demeritt, D. 1994. The Nature of Metaphors in Cultural Geography and Environmental History. *Progress in Human Geography* 18: 163–85.

Desai, M. 2012. In Search of the Sacred and Antique in Colonial India. In *Colonial Frames, Nationalist Histories*, ed. M. Rajagopalan and M. Desai, 47–71. Burlington, VT: Ashgate Publishing.

D'Souza, R. 2002. Colonialism, Capitalism, and Nature: Debating the Origins of Mahanandi Delta's Hydraulic Crisis (1803–1928). *Economic and Political Weekly* 37: 1261.

_____. 2006. Water in British India: the Making of a "Colonial Hydrology." *Compass* 4: 621.

D'Souza, R., P. Mukhopadhyay, and A. Kothari. 1998. Re-evaluating Multi-purpose River Valley Projects: a Case Study of Hirakud, Ukai, and IGNP. *Economic and Political Weekly* 33: 6.

Doctor-Pingel, M. 2012. Poppo Pingel, Auroville Architects Monograph Series. Ahmedabad: Mapin.

Doshi, S. 2013. Resettlement Ecologies: Environmental Subjectivity and Graduated Citizenship in Mumbai. In *Ecologies of Urbanism in India: Metropolitan Civility and Sustainability*, ed. A. Rademacher and K. Sivaramakrishnan, 225–48. Hong Kong: Hong Kong University Press.

Dossal, M. 1991. *Imperial Designs and Indian Realities: the Planning of Bombay City, 1845–1875*. London: Oxford University Press.

Dove, M.R. 2001. Inter-Disciplinary Borrowing in Environmental Anthropology and the Critique of Modern Science. In *New Directions in Anthropology and Environment: Intersections*, ed. C. L. Crumley, 90–110. Walnut Creek, CA: AltaMira Press.

Durand-Lasserve, A. 2006. Informal settlements and the Millennium Development Goals: Global Policy Debates on Property Ownership and Security of Tenure. *Global Urban Development* 2 (1): 1–15.

Economist Intelligence Unit. 2011. *Asian Green City Index: Assessing the Performance of Asia's Major Cities*. Munich: Siemens.

Edwards, A. 2005. *The Sustainability Revolution: Portrait of a Paradigm Shift*. New York: New Society Press.

Elkington, J. 2001. *The Triple Bottom Line: Does it All Add Up?* London: Earthscan.

Ernstson, H. 2013. The Social Production of Ecosystem Services: A Framework for Studying Environmental Justice and Ecological Complexity in Urbanized Landscapes. *Landscape and Urban Planning* 109 (1): 7–17.

Evenson, N. 1989. *The Indian Metropolis: A View Toward the West*. New Haven: Yale University Press.

Farber, S. et. al. 2006. Linking Ecology and Economics for Ecosystem Management. *Bioscience* 56 (2): 117–29.

Faure, D., and H. F. Siu. 1995. *Down to Earth: the Territorial Bond in South China*. Stanford: Stanford University Press.

Fennell, C. 2015. "Emplacement" for "Theorizing the Contemporary." *Cultural Anthropology*. https://culanth.org/conversations/17-theorizing-the-contemporary.

Fischer, M. J. 2007. Four Genealogies for a Recombinant Anthropology of Science and Technology. *Cultural Anthropology* 22 (4): 539–615.

Freitag, S. B. 1980. Sacred Symbol as Mobilizing Ideology: The North Indian Search for a "Hindu" community. *Comparative Studies in Society and History* 4: 597–625.

Foucault, M. 1984. Space, Knowledge, and Power. In *Foucault Reader*, ed. P. Rabinow, 239–56. New York: Pantheon Books.

Franklin, S. 1995. Science of Cultures, Cultures of Science. *Annual Review of Anthropology* 24: 162–85.

Fuchs, R. 2010. Cities at Risk: Asia's Coastal Cities in an Era of Climate Change. *Asia Pacific Issues* 96.

Gandy, M. 2010. Vicissitudes of Urban Nature: Transitions and Transformations at a Global Scale. *Radical History Review* 107: 178–84.

Ghassam-Fachandi, P. 2012. *Pogrom in Gujarat: Hindu Nationalism and Anti-Muslim Violence in India*. Princeton: Princeton University Press.

Ghertner, A. 2013. Nuisance Talk: Middle Class Discourses of a Slum Free Delhi. In *Ecologies of Urbanism in India: Metropolitan Civility and Sustainability*, ed. A. Rademacher and K. Sivaramakrishnan, 249–76. Hong Kong: Hong Kong University Press.

Giardet, H. 1993. *The Gaia Atlas of Cities: New Directions for Sustainable Urban Living*. New York: Anchor Books.

Gieryn, T. 1995. The Boundaries of Science. In *Handbook of Science and Technology Studies*, ed. S. Jasanoff et al. Thousands Oaks, CA: Sage.

Gissen, D. 2003. *Big and Green: Toward Sustainable Architecture in the 21st Century*. Princeton: Princeton Architectural Press.

Glassie, H. 1975. *Folk Housing in Middle Virginia*. Knoxville, TN: University of Tennessee Press.

Gooding, D. 1992. Putting Agency Back into Experiment. In *Science as Practice and Culture*, ed. A. Pickering, 65–112. Chicago: University of Chicago Press.

Guhathakurta, S., ed. 2003. *Integrated Land Use and Environmental Models*. Berlin: Springer.

Guha, R., and Martinez-Alier, J. 1997. *Varieties of Environmentalism: Essays North and South*. London: Earthscan.

Günel, G. Forthcoming. *Spaceship in the Desert: Energy, Climate Change and Green Business in Abu Dhabi*. Durham, NC: Duke University Press.

_____. 2016. Inhabiting the Spaceship: The Connected Isolation of Masdar City. In *Climates: Architecture and the Planetary Imaginary*, ed. J. Graham, 361–73. New York: Columbia Books on Architecture and the City.

Gupta, P.K. 2000. *Methods in Environmental Analysis*. Delhi: Agrobios India.

Gurevitch, J., and D. K. Padilla. 2004. Are Invasive Species a Major Cause of Extinctions? *Trends in Ecology and Evolution* 19 (9): 470–74.

Hall, P. 2002. *Cities of Tomorrow: An Intellectual History of Urban Planning and Design in the Twentieth Century*. New York: Wiley-Blackwell.

Hansen, T. B. 2001. *Wages of Violence: Naming and Identity in Postcolonial Bombay*. Princeton: Princeton University Press.

_____. 1999. *The Saffron Wave: Democracy and Hindu Nationalism in Modern India*. Princeton: Princeton University Press.

Hansen, T. B., and O. Varkaaik. 2009. Urban Charisma: On Everyday Mythologies in the City. *Critique of Anthropology* 29 (1): 5–26.

Harms, E. 2011. *Saigon's Edge: On the Margins of Ho Chi Minh City*. Minneapolis: University of Minnesota Press.

Haraway, Donna. 1989. *Primate Visions: Gender, Race, and Nature in the World of Modern Science*. New York: Routledge.

_____. 1991. *Simians, Cyborgs, and Women: the Reinvention of Nature*. New York: Routledge.

_____. 1997. *Modest Witness@Second Millenium. FemaleMan-Meets-OncoMouse: Feminism and Technoscience*. New York: Routledge.

Harvey, D. 1990. *The Condition of Postmodernity: an Enquiry into the Origins of Cultural Change*. Cambridge: Blackwell.

_____. 2008. The Right to the City. *New Left Review* 53.

_____. 1973. *Social Justice and the City*. Baltimore: Johns Hopkins University Press.

_____. 2000. *Spaces of Hope*. Berkeley: University of California Press.

_____. 1999. Considerations on the Environment of Justice. In *Global Ethics and the Environment*, ed. N. Low. London: Routledge.

_____. 2010. *The Enigma of Capital and the Crises of Capitalism*. New York: Oxford University Press.

Hassan, P. 2001. Marking Identity through Vernacular Form in Bengal. In *Traditional and Vernacular Architecture: Proceedings of the Seminar*, ed. S. Krishnaswamy, 31–39. Chennai: Madras Craft Foundation.

Hernandez, C., and R. Mayur. 2010. *Pedagogy of the Earth* and *Green Living Design for a Sustainable Earth*. Mumbai: International Institute for a Sustainable Future/Green Books Press.

Hetherington, K. 2014. Waiting for the Surveyor: Development Promises and the Temporality of Infrastructure. *Journal of Latin American and Caribbean Anthropology* 19 (2): 195–211.

Heynen, N. 2003. The Scalar Production of Injustice Within the Urban Forest. *Antipode* 35 (5): 980–98.

Hilgartner, S. 2000. *Science on Stage: Expert Advice as Public Drama*. Stanford: Stanford University Press.

Hillier, W., and J. Hanson. 1984. *The Social Logic of Space*. Cambridge: Cambridge University Press.

Hosagrahar, J. 2005 *Indigenous Modernities: Negotiating Architecture and Urbanism*. New York: Routlege.

Hobsbawm, E., and T. Ranger. 1983. *The Invention of Tradition*. Cambridge: Cambridge University Press.

Humphrey, C. 2005. Ideology in Infrastructure: Architecture and Soviet Imagination. *Journal of the Royal Anthropological Institute* 11 (1): 39–58.

Ingersoll, R. 1996. Second Nature: On the Social Bond of Ecology and Architecture. In *Reconstructing Architecture: Critical Discourses and Social Practices*, ed. T. Dutton and L. Hurst Mann, 119–57. Minneapolis: University of Minnesota Press.

Jacobs, J. 2002. *The Nature of Economies*. New York: Random House.

Jasanoff, S. 2003. (No?) Accounting for Expertise. *Science and Public Policy* 30 (30): 157–62.

_____. 2004a. Ordering Knowledge, Ordering Society. In *States of Knowledge: The Co-production of Science and Social Order*. New York: Routledge.

_____, ed. 2004b. *States of Knowledge: The Co-production of Science and Social Order*. New York: Routledge.

Jaffrelot, Christophe. 1998. *The Hindu Nationalist Movement in India*. New York: Columbia University Press.

_____. 1993. Hindu Nationalism: Strategic Syncretism in Ideology Building. *Economic and Political Weekly* 28 (12/13): 517–24.

Karkaria, B. 2014. Mumbai is on the Verge of Imploding. *The Guardian*, November 23, 2014. http://www.theguardian.com/cities/2014/nov/24/mumbai-verge-imploding-polluted-megacity.

_____. 2007. Lose the Vultures, Lose the Soul. *New York Times*, May 11, 2007.

Kidambi, P. 2007. *The Making of an Indian Metropolis: Colonial Governance and Public Culture in Bombay, 1890–1920*. Ashgate: Aldershot.

King, A. 2004. *Spaces of Global Cultures: Architecture, Urbanism, Identity*. New York: Routledge

_____. 1990. *Urbanism, Colonialism, and the World Economy*. New York: Routledge.

Kumar, H. D. 1992. *Modern Concepts of Ecology*. Delhi: Vikas Publishing House.

Latour, B. 1993. *We Have Never Been Modern*. Trans. Catherine Porter. Cambridge: Harvard University Press.

_____. 1988. *Science in Action: How to Follow Scientists and Engineers through Society*. Cambridge: Harvard University Press.

Lawhon, M., H. Ernstson, and J. Silver. 2014. Provincializing Urban Political Ecology: Towards a Situated UPE through African Urbanism. *Antipode* 46 (2): 497–516.

Lawrence, D., and S. Low. 1990. The Built Environment and Spatial Form. *Annual Review of Anthropology* 19: 453–505.

Leach, N. 1997. *Rethinking Architecture: A Reader in Critical Theory*. London: Routledge.

Lefebvre, H. 1991. *The Production of Space*. Cambridge: Blackwell.

_____. 1970/2003. *The Urban Revolution.* Minneapolis: University of Minnesota Press.

Levi Strauss, C. 1966. *The Savage Mind.* London: Weidenfeld and Nicolson.

Lewis, M. 2004. *Inventing Global Ecology: Tracking the Biodiversity Ideal in India 1947–1997.* Athens, OH: Ohio University Press.

Lodge, D.M. 1993. Biological Invasions: Lessons for Ecology. *Trends in Ecology and Evolution* 8 (4): 133–37.

Lovelock, J. 1979. *Gaia: A New Look at Life on Earth.* London: Oxford University Press.

Lynch, M. 1985. *Art and Artifact in Laboratory Science: A Study of Shop Work and Shop Talk in the Laboratory.* London: Routledge.

Mack, R. N. et al. 2000. Biotic Invasions: Causes, Epidemiology, Global Consequences, and Control. *Issues in Ecology* 5.

Mathews, A. 2011. *Instituting Nature: Authority, Expertise, and Power in Mexican Forests.* Cambridge: MIT Press.

Martinez-Alier, J., 2004. Political Ecology, Distributional Conflicts, and Economic Incommensurability. *Sustainability: Sustainable Development* 3: 395–415.

Mishra, P. 2006. Ayodhya: The Modernity of Hinduism. In *Temptations of the West: How to Be Modern in India, Pakistan, Tibet and Beyond,* 80–110. New York: Farrar, Straus and Giroux.

Marx, K. 1887/1967. Vol. 1 of *Capital: A Critique of Political Economy.* New York: International Publishers.

McFarlane, C. 2008. Governing the Contaminated City: Infrastructure and Sanitation in Colonial and Post-colonial Bombay. *International Journal of Urban and Regional Research* 32: 415–35.

McDonough, W., and M. Braungart. 2002. *Cradle to Cradle: Remaking the Way We Make Things.* New York: North Point Press.

McKay, R. 2012. Afterlives: Humanitarian Histories and Critical Subjects in Mozambique. *Cultural Anthropology* 27 (2): 286–309.

McKinsey Global Institute. 2009. *Promoting Energy Efficiency in the Developing World.* Report.

_____. 2010. *India's Urban Awakening: Building Inclusive Cities, Sustaining Economic Growth.* Report.

McLeod, M. 1983. Architecture or Revolution: Taylorism, Technocracy, and Social Change. *Art Journal* 43 (2): 132–14.

Mehrotra, R. 2004. Constructing Cultural Significance: Looking at Bombay's Historic Fort Area. *Future Anterior: Journal of Historic Preservation, History, Theory, and Criticism* 1 (2): 24–31.

Meister, M. 2001. Vernacular Architecture and the Rhetoric of Re-making. In *Traditional and Vernacular Architecture: Proceedings of the Seminar,* ed. S. Krishnaswamy, 9–15. Chennai: Madras Craft Foundation.

Merry, S., K. Davis, and B. Kingsbury, eds. 2015. *The Quiet Power of Indicators: Measuring Governance, Corruption, and the Rule of Law.* Cambridge: Cambridge University Press.

Metcalf, T. R. 1989. *An Imperial Vision: Indian Architecture and Britain's Raj.* Berkeley: University of California Press.

Mitchell, T. 2002. *Rule of Experts.* Berkeley: University of California Press.

_____. 2002. Can the Mosquito Speak? In *Rule of Experts: Egypt, Technopolitics, Modernity*. Berkeley: University of California Press.

Morgan, J., and A. Richards. 1990. *A Paradise Out of the Common Field: The Pleasures and Plenty of the Victorian Garden*. New York: Harper and Row.

Mumbai Waterfronts Center and PK Das & Associates. 2012. *Open Mumbai: Re-envisioning the city and its open spaces*. Mumbai: PK Das and Associates. Full-length exhibition book.

Murphy, K. 2015. *Swedish Design: an Ethnography*. Ithaca: Cornell University Press.

_____. 2016. Design as a Cultural System. *Epic*, May 16. https://www.epicpeople.org/design-as-a-cultural-system/.

Nebel, B.J., and R.T. Wright. 1999. *Environmental Science: The Way the World Works*. New York: Prentice Hall College Division.

Nijman, J. 2008. Against the Odds: Slum Rehabilitation in Neoliberal Mumbai. *Cities* 25: 73–85.

Odum, E. 1983. *Basics of Ecology*. New York: Harcourt Brace College Publishers.

Ogden, L., B. Hall, and K. Tanita. 2010. Animals, Plants, People, and Things: a Review of Multispecies Ethnography. *Environmental Sociology* 4: 5–24.

Ong A. 1999. *Flexible Citizenship: The Cultural Logics of Transnationality*. Durham: Duke University Press.

Orr, Y., J. S. Lancing, and M. R. Dove. 2015. Environmental Anthropology: Systemic Perspectives. *Annual Review of Anthropology* 44: 153–68.

Pandey, G. 1983. Rallying Round the Cow: Sectarian Strife in the Bhojpuri Region, c. 1888–1917. In Vol. 2 of *Subaltern Studies*, ed. R. Guha. Oxford: Oxford University Press.

Pandian, A., and D. Ali, eds. 2010. *Ethical Life in South Asia*. Bloomington: Indiana University Press.

Paniker, S. 2008. Indian Architecture and the Production of a Postcolonial Discourse: A Study of Architecture and Design 1984–1992. PhD diss., University of Adelaide.

Patankar, A. et al. 2010. Mumbai City Report. *International Workshop on Climate Change Vulnerability Assessment and Urban Development Planning for Asian Coastal Cities*. Report.

Payne, G. 1997. *Urban Land Tenure and Property Rights in Developing Countries: A Review*. London: Practical Action.

Peet, R., and M. Watts, eds. 1996. *Liberation Ecologies: Environment, Development, Social Movements*. New York: Routledge.

Pickett, S. T. A., and P.S. White, eds. 1985. *The Ecology of Natural Disturbance and Patch Dynamics*. London: Academic Press.

Pickett, S. T. A. et. al. 1997. A Conceptual Framework for the Study of Ecosystems in Urban Areas. *Urban Ecosystems* 1: 185–99.

Pickett, S. T. A. et al. 2001. Urban Ecological Systems: Linking Terrestrial, Ecological, Physical, and Socioeconomic Components of Metropolitan Areas. *Annual Review of Ecology and Systematics* 32: 127–57.

Pickett, S.T.A., and M.L. Cadenasso. 2002. The Ecosystem as a Multidimensional Concept: Meaning, Model, and Metaphor. *Ecosystems* 5: 1–10.

Pickett, S. T. A., M. L. Cadenasso, and B. McGrath, eds. 2013. *Resilience in Ecology and Urban Design: Linking Theory and Practice for Sustainable Cities*. New York: Springer.

Prakash, G. 2010. *Mumbai Fables.* Princeton: Princeton University Press.

Rademacher, A. 2009. When is Housing an Environmental Problem? Reforming Informality in Kathmandu. *Current Anthropology* 50 (4): 513–34.

_____. 2011. *Reigning the River: Urban Ecology and Political Transformation in Kathmandu.* Durham: Duke University Press.

_____. Urban Political Ecology. *Annual Review of Anthropology* 44: 137–52.

Rademacher, A., and K. Sivaramakrishnan, eds. 2017. *Places of Nature in Ecologies of Urbanism.* Hong Kong: Hong Kong University Press.

_____, eds. 2013. *Ecologies of Urbanism in India: Metropolitan Civility and Sustainability.* Hong Kong: Hong Kong University Press.

Rajagopalan, M. 2012. From Colonial Memorial to National Monument: The Case of the Kashmiri Gate, Delhi. In *Colonial Frames, Nationalist Histories: Imperial Legacies, Architecture, and Modernity,* ed. M. Rajagopalan and M. Desai. Surrey, UK: Ashgate.

_____. 2011. A Medieval Monument and its Modern Myths of Iconoclasm: The Enduring Contestations over the Qutb Complex in Delhi, India. In *Reuse Value: Spoliation and Appropriation in Art and Architecture from Constantine to Sherrie Levine,* ed. R. Brilliant and D. Kinney. Surrey, UK: Ashgate.

Rajagopalan, M., and M. Desai, eds. 2012. *Colonial Frames, Nationalist Histories: Imperial Legacies, Architecture, and Modernity.* Surrey, UK: Ashgate.

Rao, N. 2013. *House, but no Garden: Apartment Living in Bombay's Suburbs, 1898–1964.* Minneapolis: University of Minnesota Press.

Rao, V. 2011. A New Urban Type: Gangsters, Terrorists, Global Cities. *Critique of Anthropology* 31: 1.

_____. 2008. Pirating Utopia: Mumbai at the End of the Planning Era. In *Mumbai Reader 07,* ed. P. Joshi, I. Mathew, I. Nazareth, and R. Mehotra. Mumbai: Urban Design Research Institute.

Rawal, R. et al. 2012. Energy Code Enforcement for Beginners: A Tiered Approach to Energy Code in India. *ACEEE Summer Study on Energy Efficiency in Buildings.* Report.

Rebele, F. 1994. Urban Ecology and Special Features of Urban Ecosystems. *Global Ecology and Biogeography Letters* 4: 173–87.

Rogers, R. 1997. *Cities for a Small Planet.* New York: Westview.

Roy, A. 2009. Civic Governmentality: The Politics of Inclusion in Beirut and Mumbai. *Antipode* 41: 159–79.

Roy, A., and A. Ong, eds. 2012. *Worlding Cities: Asian Experiments and the Art of Being Global.* London: Wiley-Blackwell.

Ryan, J. C. 2012. Passive Flora? Reconsidering Nature's Agency through Human-Plant Studies. *Societies* 2: 101–21.

Sanyal, B., and V. Mukhija. 2001. Institutional Pluralism and Housing Delivery: A Case of Unforeseen Conflicts in Mumbai, India. *World Development* 29: 2043–57.

Sassen, S. 1991. *The Global City: New York, London, Tokyo.* Princeton: Princeton University Press.

Sears, T. I. 2001. Whither Vernacular Architecture? In *Traditional and Vernacular Architecture: Proceedings of the Seminar,* ed. S Krishnaswamy, 133–40. Chennai: Madras Craft Foundation.

Shanks, M., and C. Tilley. 1987. *Social Theory and Archaeology.* Oxford: Polity Press.

_____. 1992. *Re-constructing Archaeology*. London: Routledge.

Sharan, A. 2014. *In the City, Out of Place: Nuisance, Pollution, and Dwelling in Delhi, c. 1850–2000*. Delhi: Oxford University Press.

Sharma, K. 2000. *Rediscovering Dharavi*. New Delhi: Penguin.

Singer, M. 2014. Zoonotic Ecosyndemics and Multispecies Ethnography. *Anthropological Quarterly* 87 (4): 1279–1309.

Singh, M.K. 2010. City of Dreams? *Times of India*, November 15, 2010.

Simone, A. 2004. *For the City yet to Come: Changing African Life in Four Cities*. Durham: Duke University Press.

Sivaramakrishnan, K. 1999. *Modern Forests: Statemaking and Environmental Change in Colonial Eastern India*. Stanford: Stanford University Press.

_____. 2015. Ethics of Nature in Indian Environmental History: A Review Article. *Modern Asian Studies* 49 (4): 1261–1310.

Srinivas, S. 2015. *A Place for Utopia: Urban Designs from South Asia*. Seattle: University of Washington Press.

Swyngedouw, E. 1996. The City as a Hybrid: on Nature, Society, and Cyborg Urbanisation. *Capitalism, Nature, Socialism* 7 (2): 65–80.

Tarlo, E. 2000. Welcome to History: A Resettlement Colony in the Making. In *Delhi: Urban Space and Human Destinies*, ed. V. Dupont, E. Tarlo, and D. Vidal, 51–74. New Delhi: Manohar.

Taylor, P., and F. Buttel. 1992. How Do We Know We Have Global Environmental Problems? Science and the Globalization of Environmental Discourse. *Geoforum* 23 (3): 405–16.

Turner, B. L. II, and P. Robbins. 2008. Land Change Science and Political Ecology: Similarities, Differences, and Implications for Sustainability Science. *Annual Review of Environmental Resources* 33: 295–316.

Tsing, A. 2000. The Global Situation. *Current Anthropology* 15: 327–60.

_____. 2001. Nature in the Making. In *New Directions in Anthropology and Environment*, ed. C. L. Crumley, 3–23. Lanham, Maryland: Altamira Press.

_____. 2005. *Friction: An Ethnography of Global Connections*. Princeton: Princeton University Press.

_____. 2012. Unruly Edges: Mushrooms as Companion Species. *Environmental Humanities* 1: 141–54.

United Nations. 2011. *Revision, World Urbanization Prospects*. Report.

Udyavar, R., and P. Shah, eds. 2010. *Survival at Stake: An Anthology of Essays by Rashmi Mayur*. Ahmedabad: Gujarat Institute of Civil Engineers and Architects.

Villiers-Stuart, C. 1913. *Gardens of the Great Mughals*. London: Adam and Charles Black.

Weinstein, L. 2014. *Durable Slum: Dharavi and the Right to Stay Put in Globalizing Mumbai*. Minneapolis: University of Minnesota Press.

_____. 2008. Mumbai's Development Mafias: Globalization, Organized Crime, and Land Development. *IJURR* 32: 22–39.

Whatmore, S. 1999. Hybrid Geographies: Rethinking the Human in Human Geography. In *Human Geography Today*, ed. D. Massey, J. Allen, and P. Sarre, 22–40. Cambridge: Polity.

_____. 2003. Generating Materials. In *Using Social Theory: Thinking through Research*, ed. M. Pryke, S. Rose, and S. Whatmore, 89–104. London: Sage.

Williamson, T. J. 2002. *Understanding Sustainable Architecture*. New York: Taylor & Francis.

Wolf, E. 1982. *Europe and the People without History.* Berkeley: University of California Press.

Worster, D. 1990. The Ecology of Order and Chaos. *Environmental History Review* 14: 1–18.

_____. 1994. *Nature's Economy: A History of Ecological Ideas.* London: Cambridge University Press.

_____. 1996. The Two Cultures Revisited. *Environment and History* 2 (1): 3–14.

Zeiderman, A. 2016. *Endangered City: The Politics of Security and Risk in Bogotá.* Durham: Duke University Press.

Zimmerer, K. 2000. The Reworking of Conservation Geographies: Nonequilibrium Landscapes and Nature Society Hybrids. *Annals of the Association of American Geographers* 90 (2): 356–69.

INDEX

197